BURLEIGH DODDS SCIENCE: INSTANT INSIGHTS

NUMBER 104

# Agroforestry practices

Published by Burleigh Dodds Science Publishing Limited
82 High Street, Sawston, Cambridge CB22 3HJ, UK
www.bdspublishing.com

Burleigh Dodds Science Publishing, 1518 Walnut Street, Suite 900, Philadelphia, PA 19102-3406, USA

First published 2024 by Burleigh Dodds Science Publishing Limited


British Library Cataloguing in Publication Data
A catalogue record for this book is available from the British Library

ISBN 978-1-80146-989-0 (Print)
ISBN 978-1-80146-990-6 (ePub)

DOI: 10.19103/9781801469906

Typeset by Deanta Global Publishing Services, Dublin, Ireland

# Contents

# Series list

| Title | Series number |
|---|---|
| Sweetpotato | 01 |
| Fusarium in cereals | 02 |
| Vertical farming in horticulture | 03 |
| Nutraceuticals in fruit and vegetables | 04 |
| Climate change, insect pests and invasive species | 05 |
| Metabolic disorders in dairy cattle | 06 |
| Mastitis in dairy cattle | 07 |
| Heat stress in dairy cattle | 08 |
| African swine fever | 09 |
| Pesticide residues in agriculture | 10 |
| Fruit losses and waste | 11 |
| Improving crop nutrient use efficiency | 12 |
| Antibiotics in poultry production | 13 |
| Bone health in poultry | 14 |
| Feather-pecking in poultry | 15 |
| Environmental impact of livestock production | 16 |
| Sensor technologies in livestock monitoring | 17 |
| Improving piglet welfare | 18 |
| Crop biofortification | 19 |
| Crop rotations | 20 |
| Cover crops | 21 |
| Plant growth-promoting rhizobacteria | 22 |
| Arbuscular mycorrhizal fungi | 23 |
| Nematode pests in agriculture | 24 |
| Drought-resistant crops | 25 |
| Advances in detecting and forecasting crop pests and diseases | 26 |
| Mycotoxin detection and control | 27 |
| Mite pests in agriculture | 28 |
| Supporting cereal production in sub-Saharan Africa | 29 |
| Lameness in dairy cattle | 30 |
| Infertility and other reproductive disorders in dairy cattle | 31 |
| Alternatives to antibiotics in pig production | 32 |
| Integrated crop-livestock systems | 33 |
| Genetic modification of crops | 34 |

# Acknowledgements

Chapters in this Instant Insight are taken from the following sources:

Chapter 1 Integrated crop-livestock systems with agroforestry to improve organic animal farming

Chapter taken from: Vaarst, M. & Roderick, S. (ed.), Improving organic animal farming, Burleigh Dodds Science Publishing, Cambridge, UK, 2019, (ISBN: 978 1 78676 180 4; www.bdspublishing.com)

Chapter 2 Tree planting and management in agroforestry

Chapter taken from: Mosquera-Losada, M. R. and Prabhu, R. (ed.), Agroforestry for sustainable agriculture, Burleigh Dodds Science Publishing, Cambridge, UK, 2019, (ISBN: 978 1 78676 220 7; www.bdspublishing.com)

Chapter 3 Temperate alley cropping systems

Chapter taken from: Mosquera-Losada, M. R. and Prabhu, R. (ed.), Agroforestry for sustainable agriculture, Burleigh Dodds Science Publishing, Cambridge, UK, 2019, (ISBN: 978 1 78676 220 7; www.bdspublishing.com)

Chapter 4 Sustainable production of willow for biofuel use

Chapter taken from: Saffron, C. (ed.), Achieving carbon-negative bioenergy systems from plant materials, Burleigh Dodds Science Publishing, Cambridge, UK, 2020, (ISBN: 978 1 78676 252 8; www.bdspublishing.com)

Chapter 5 The importance of agroforestry systems in supporting biodiversity conservation and agricultural production: a European perspective

Chapter taken from: Saffron, C. (ed.), Achieving carbon-negative bioenergy systems from plant materials, Burleigh Dodds Science Publishing, Cambridge, UK, 2020, (ISBN: 978 1 78676 252 8; www.bdspublishing.com)

# Chapter 1

## Integrated crop-livestock systems with agroforestry to improve organic animal farming

*A. J. Escribano, Nutrion Internacional, Spain; J. Ryschawy, University of Toulouse, France; and L. K. Whistance, The Organic Research Centre, UK*



## 1 Introduction

The International Federation of Organic Movements (IFOAM) has established four fundamental principles which underpin organic farming (IFOAM, 2005; Luttikholt, 2007):

- Health: organic agriculture should sustain and enhance the health of soil, plant, animal, human and planet
- Ecology: organic agriculture should be based on living ecological systems and cycles, work with them, emulate and help to sustain them
- Fairness: organic agriculture should build on relationships that ensure fairness with regard to the common environment and life opportunities
- Care: organic agriculture should be managed in a precautionary and responsible manner to protect the health and well-being of current and future generations and the environment

http://dx.doi.org/10.19103/AS.2017.0028.12

Based on these principles, organic farm standards are designed to reduce the environmental impact of agriculture, to encourage socially and ethically just food production and to produce safe and healthy food. In organic animal production, the following are included (Sundrum, 2001):

- Maximum use of natural resources (e.g. pasture) and self-sufficiency in resource use to minimise environmental impact
- High animal welfare standards (e.g. lower stocking densities, more access to the outdoors for livestock and other ways of encouraging natural behaviour)
- Reduced reliance on veterinary interventions

One way of achieving these principles is the use of integrated crop-livestock systems (ICLS), which is very common for well-balanced organic systems with animals. In some parts of the world, complex ICLS do not have any formal organic certification, in part because the costs of certification may not be recouped by any price premium from organic labelling. However, these systems do conform to IFOAM principles and, in some ways, could even be seen as fulfilling those principles more effectively than some organic monocultural dairy or beef production systems. It has been argued that organic systems involving grassland and livestock may be the most sustainable form of organic production (Younie, 2000). However, farming systems with trees may have greater potential to improve sustainability in terms of being climate resilient and being part of landscapes which are 'climate friendly' (sequester carbon), and by improving or maintaining biodiversity in terms of plants, insects and wildlife. More broadly, ICLS - including agroforestry as an interesting case - have been seen as a key form of ecological intensification to achieve both food security and environmental sustainability (Lemaire et al., 2014). This chapter reviews types of such systems, with a focus on agroforestry systems, their potential environmental and economic benefits, their contributions to animal health and welfare, as well as challenges in implementing and managing such systems whilst ensuring a high animal health and welfare level.

## 2 Types of ICLS

One way of achieving sustainable land management is the implementation of diversified systems combining crops, grasslands and livestock (Smith et al., 2012). These diversified systems can also include agroforestry (Nair et al., 2009). Interest in these more diversified and integrated systems of agricultural production can be seen, in part, as a reaction to the environmental and other problems associated with modern, highly specialised and more intensive forms of animal production. In these systems, for example, crops for

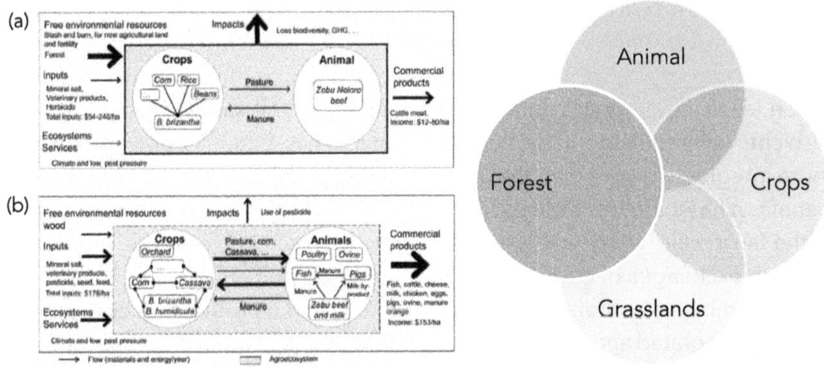


**Figure 1** Diversification and interconnection of components in a crop-livestock system including forest, crops, livestock and grasslands (Bonaudo et al., 2014), and the adaptation of the framework of Moraine et al. (2014) integrated crop-livestock-forest systems.

animal feed may be grown in separate locations and transported significant distances to meet the needs of large, monocultural dairy or beef farms which are effectively decoupled from local production (Vaarst, 2015). Opportunities to integrate (or re-integrate) crops and livestock thus exist not only in developing countries but also in developed countries such as the United States (Franzluebbers, 2007; Sulc and Tracy, 2007). These systems are known as ICLS or, where they include agroforestry, integrated crop-livestock-forestry systems (ICLFS) (Jose, 2009; Kremen et al., 2012; Kremen and Miles, 2012; Smith et al., 2012; Cook and Ma, 2014; Moraine et al., 2014). These systems are illustrated in Fig. 1.

In an integrated system, livestock and crops are produced within a coordinated framework (Van Keulen and Schiere, 2004; IFAD, 2010). The waste products of one component serve as a resource for the other: manure is used to fertilise soil for crops whilst crop residues and by-products are used to feed animals. There is a large diversity within ICLS as highlighted by Ryschawy et al. (2012) in France or Russelle et al. (2007) in the United States. The ways the different components in such systems integrate influences the environmental and economic effects of such systems (Ryschawy et al., 2014a,b). Moraine et al. (2014, 2016) have sought to conceptualise different levels of integration between crops, grasslands and livestock in ICLS. They have identified three main types of integration:

- local co-existence
- complementarity between components in one location (e.g. use of crops/grasslands to feed animals, use of animal manure to fertilise soils)
- real synergy in which the system is achieving self-sufficiency in inputs

These forms of integration correspond to a gradient of spatial, temporal and social coordination between components (livestock, crops and grasslands) which, when combined, promote a range of environmental and other benefits. These levels of integration are illustrated in Fig. 2. Influenced by the concept of niche differentiation in ecological theory, it has been suggested that the more integrated a system is, the greater the potential environmental and other benefits it has to deliver (Bonaudo et al., 2014). Such systems can be considered at the farm scale, which is the most common form of ICLS, or at the local level, through exchanges of crops and manure between farmers.

A number of authors argue that many so-called crop–livestock systems are not truly integrated and that, in particular, there has been a significant decoupling of livestock from the other elements in many agricultural systems (Escribano et al., 2014; Moraine et al., 2014). An example of local co-existence in ICLS highlighted by Perrot et al. (2008) are dairy-crops systems in the French region of Lorraine. In these systems large high-input dairy units operate alongside high-input cash crop systems, with manure from the dairy farms used on the crops and maize silage used to supplement the cows' diet. Although integration is limited in these co-existence systems, profitability can be higher than truly ICLS, in part because of subsidies (Veysset at al., 2005). Ryschawy et al. (2012) has highlighted the environmental and economic benefits of exchanges of crop and livestock products between conventional and organic farmers in south-western France. Lemaire et al. (2014)

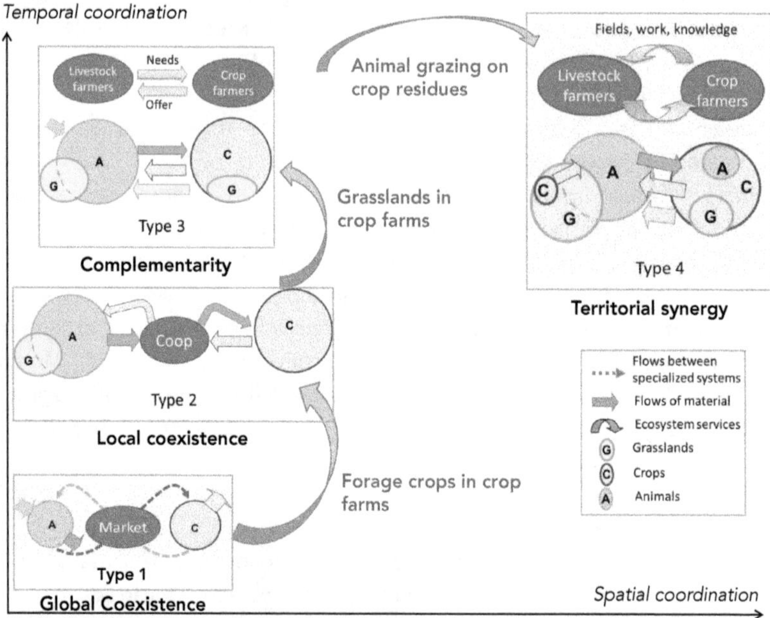


**Figure 2** Levels of crop-livestock integration at farm level and beyond. Adapted from: Moraine et al. (2014).

have highlighted the ecological benefits of greater collaboration and integration between farms in a particular area.

At the other end of the scale is the 'dehesa' ICLFS found in Spain and Portugal. This is a multipurpose production system combining crops, livestock and forestry where different agricultural uses (cork, firewood, crops etc.) have been traditionally combined with different livestock species (sheep, pigs, cattle etc.) (Moreno et al., 2007). The integration of different livestock species, for example, allows an optimal use of resources by avoiding competition for food due to animals' different grazing behaviours (Anderson et al., 2012a,b).

## 3 Environmental and economic benefits of ICLS

ICLS have been seen as a way to address agricultural sustainability issues. The integration of domesticated animals into farming systems can give long-term benefits in circulation of nutrients, utilisation and care of land areas, prevention of land degradation and erosion, and contribution to resilient and sustainable farming systems (Bonaudo et al., 2014). A number of authors suggest that synergies between crops and livestock can improve nutrient cycling and delivery of ecosystem services in agricultural systems (Ryschawy et al., 2014b; Behera et al., 2015; Moraine et al., 2016). As an example, it has been suggested that they can increase carbon sequestration, improve soil quality, reduce pollution risks (e.g. by reducing the need for fertiliser) and reduce fire risks (Plieninger, 2007; Tárrega et al., 2009; Henkin et al., 2011; Núñez et al., 2012; Dumont et al., 2013; Riedel et al., 2013; Sanderson et al., 2013; Simón et al., 2013; Soussana and Lemaire, 2014). Studies have found soil properties (soil bulk density, volumetric soil-water content, soil penetration resistance, total porosity, macro porosity, effective microporosity, unsaturated pores) are better in crop-livestock compared to monocultural systems (Acosta-Martínez et al., 2004; Marchão et al., 2007). Vilela et al. (2011) have suggested that, in addition to improving the chemical, physical and biological properties of the soil, these systems could reduce the occurrence of diseases, pests and weeds.

Looking at economic benefits, achieving greater self-sufficiency in resources in ICLS results in less dependence on external suppliers, fewer inputs, increased self-reliance and, therefore, lower production costs (Wilkins, 2008). As an example, grazed wheat and rye (*Secale cereale* L.) systems with cotton (*Gossypium hirsutum* L.) in a two-paddock rotation were found to be less dependent on irrigation and chemical inputs, using 23–25% less irrigation water per hectare, as well as 40% less N fertiliser, with improved soil microbial and enzymatic activities, enhanced C sequestration and greater rainfall infiltration than monoculture systems (Allen et al., 2005, 2007). Martha et al. (2011) have identified the potential economies of 'scope' in these systems that is cost reduction associated with producing multiple outputs, as well as the risk-reducing effects of diversification identified by others.

As well as reducing costs, it has been suggested that the diversification inherent in ICLS also diversifies sources of income, thus reducing farms' economic risk and increasing their stability, adaptability and resilience in the face of market, political and/or environmental changes (Dumont et al., 2013). Diversified systems such as ICLS can also involve lower overall yield variability and higher overall yields, as well as higher flexibility, adaptability and resilience (Escribano et al., 2014, 2015; Escribano, 2016b). In North Dakota, for example, net worth was nearly $9000 greater for farms with crops and beef cows compared with crops only (Anderson and Schatz, 2003). Studies by Allen et al. (2005, 2007) on integrating cotton-forage-beef cattle in Texas have shown this system to be economically viable, diversifying income while reducing expenditure in chemical inputs and water, and showing higher annual crop yields. Behera et al. (2015) have compared various models of integrated farming systems (IFS) to existing rice-wheat systems in India, suggesting that IFS showed potential to generate a greater farm income than the existing rice-wheat production systems. Research in Brazil suggests that ICLFS improved self-sufficiency, efficiency and resilience to market shocks (Bonaudo et al., 2014), ensured higher crop and animal productivity as well as reducing risk due to diversification (Vilela et al., 2011).

An example of what this can mean in practice is provided by Ryschawy et al. (2012, 2013, 2014a,b) who have analysed both the economic and environmental benefits of ICLS in farms in south-western France. They identified two ways of achieving these benefits:

- maximising autonomy in animal feeding and cropping systems (to minimise the need for external inputs)
- diversification of products sold on the market.

The potential economic and environmental benefits of these two approaches are outlined in Table 1. This research suggests that, while mixed crop-livestock farms did not always have significantly higher overall gross margins than crop or livestock-only farms, they were generally less vulnerable to fluctuations in input and output prices.

Another example is a long-term study of ICLS in Illinois (Sulc and Tracy, 2007). This depended on winter forage supply from corn residue plus rye +/− oat +/− turnip mixture, grown from November to January, though supplemental feed was needed in late winter, especially for pregnant cows. The economic return from the cattle operation in the integrated system was proved to be competitive with other conventional cattle systems used in Illinois. Total feed costs in their system in 2004 were $158 per cow (compared with the Illinois state average of $200) per cow. The authors also found cattle trampling on cropland caused some soil compaction, but were found to have minimal effect on soil properties and no detrimental effect on crop yield. Soil organic matter

**Table 1** Comparing autonomous and diversified systems (Ryschawy et al., 2014a,b)

| Case | Scenarios considered | Total gross margin (€/ha) | N balance (kg N/ha) |
|---|---|---|---|
| Autonomy | S0: Initial situation | *683* | *+6.2* |
| | S1: Clover storage Ploughing in oat-vetch | 704 | −0.3 |
| | S2: S1 + + Red clover set aside | 744 | +2.74 |
| Diversification | S0: Initial situation | *717* | *+50.3* |
| | S1: 10 heifers, 350 kg live weight, direct sale | 840 | +49.7 |
| | S2: S1 + short chain with one intermediary step | 773 | +49.7 |

Notes: Nitrogen farm-gate balance was calculated according to Simon and Le Corre (1992) method; italic values indicate the initial scenario for each type of strategy considered.

increased significantly within 5 years of conversion from corn-soybean rotation suggesting rapid building of soil organic matter in the integrated system. Winter grazing on crop fields suppressed weed biomass, as did the presence of cover crops. The authors found that cattle on winter cropland grazed near water sources, resulting in uneven redistribution of manure to fertilise soil.

## 4 Agroforestry as an ICLS

The potential economic and environmental benefits of integrated systems can be seen in the case of agroforestry. According to Bene et al. (1977)

> *the term agroforestry system refers to a land management model in which woody perennials (trees, bushes, etc.) are grown in association with herbaceous (crops, pastures) or livestock in a spatial arrangement, a rotation or both; where usually there are ecological and economic interactions between woody and other system components.*

Silvopasture, a type of agroforestry system, combines trees, pasture and livestock species in various ways, depending on the biophysical, economic, cultural and market context (Mosquera-Losada et al., 2009; Haile et al., 2010; Cubbage et al., 2012). They are multifunctional systems with the potential to achieve sustainable production and to protect ecosystem services (Smith et al., 2012).

Since tree roots extend deeper than crop plants, they are able to access nutrients and water reserves inaccessible to the latter. These nutrients are then recycled through fallen leaves on the soil surface, which leads to greater overall uptake of nutrients (Sinclair et al., 2000). The presence of trees thus allows a transfer of nutrients from the deeper soil layer to the surface, and vice versa. As a result, the content of soil organic matter increases, but the loss of nutrients to surface water is reduced (Nair et al., 2007; Crespo, 2008; Howlett et al., 2011;

Liu et al., 2012). Nitrogen cycling is also improved (Thevathasan and Gordon, 2004). As a result, trees reduce soil erosion (up to 65%) and nitrogen leaching (up to 28%). They also improve the overall physical and chemical properties of soil (Joffre and Rambal, 1988; Gallardo, 2003; Moreno et al., 2007; Wick and Tiessen, 2008). Finally, roots play a role in intercepting and filtering pollutants in soils (Jose et al., 2004; Thevathasan and Gordon, 2004; Volk et al., 2006; Borin et al., 2009; Schädler et al., 2010; Chu et al., 2010). Improved soil quality then improves carbon sequestration and landscape biodiversity (Garrity, 2004; Garrity et al., 2006; Palma et al., 2007), especially if compared to monoculture or pasture-only systems (Nair et al., 2011). This can lead to more productive soils and economic returns in the long term compared to monoculture systems (Rigueiro-Rodríguez et al., 2008; Benavides et al., 2009; Smith et al., 2013).

The broader role of forests in climate regulation has been widely discussed (Mutuo et al., 2005; Haile et al., 2010). However, it is also important to note the impact of woodland on local microclimates: temperature, wind speed and humidity are modified (Jose et al., 2004). Trees provide shade and protection against wind, preventing direct damage of crops from wind, moisture loss and erosion (Bird, 1998; Brandle et al., 2004). As a result, crop and pasture productivity, animal welfare (Jose et al., 2004; Smith et al., 2013) and livestock productivity can be improved (Mitlohner et al., 2001). This is especially important in hot and dry areas, and in fragile ecosystems, such as the 'dehesa' ecosystem (Moreno and Obrador, 2007) (Figs. 3 and 4). Although in Mediterranean areas like this there may be some competition for water and nutrients between pastures, crops and trees, this is more than offset by

**Figure 3** The *dehesa* agro-ecosystem (SW Spain and Portugal).

**Figure 4** Dehesa: some of the different uses of a traditional multipurpose agro-ecosystem. (a) Iberian pigs feeding on acorns, grass, worms and so on. (b) Cork extraction. (c) *Retinta*, the indigenous breed of cattle. (d) *Merino*, the indigenous breed of sheep (for cheese and meat). (e) Geese are able to feed on acorns, producing high-quality *foie gras*.

reduced evapotranspiration (Joffre and Rambal, 1988; Benavides et al., 2009). In addition, trees reduce surface runoff, increase infiltration and the water retention capacity of soil (Smith et al., 2013). Shade provided by trees and the interception of water by the roots reduce the effects of drought and maintain more stable moisture levels.

This integration of trees, grass and different livestock species creates biodiverse and functional systems that enable combining productivity and environmental protection (Smith et al., 2013). Agroforestry systems allow greater diversification (Jose, 2005), increasing economic stability (Dixon, 1995;

**Table 2** Summary of socio-economic and environmental services of agroforestry systems (with integration of extensive livestock farming)

| | |
|---|---|
| Environmental: Regulation | Climate regulation (carbon sequestration in extensive production systems) |
| | Biological control |
| | Regulation of water flows |
| | Maintenance of soil fertility |
| | Maintenance of pastures' quality |
| | Reduction of wildfire risk |
| | Erosion prevention |
| | Pollutants retention |
| | Recycling |
| Cultural (socio-economic) | Cultural and genetic heritage (local breeds and species), and aesthetic values: local tourism-economy |
| | Main sources of income (and even of food) in less developed areas |

Udawatta et al., 2008) and adaptability to the rapid changes occurring in the market, thus reducing risk (Escribano et al., 2015). Such systems also make it possible to implement secondary activities such as rural tourism, agro-tourism, birdwatching or hunting, which constitute important pillars for local economy, thus promoting sustainable rural development. From a cultural standpoint, these systems also contribute to the maintenance of landscapes and cultures linked to them, such as grazing, transhumance or traditional farming practices (McAdam and McEvoy, 2009). Environmentally, this protection of agricultural culture, traditional landscapes and ecosystem services and systems provides other environmental benefits, including the protection and promotion of biodiversity. Studies have demonstrated that the establishment of silvopastoral systems have had a positive impact on biodiversity due to the creation of habitats and corridors for beneficial fauna (Schmidt and Tscharntke, 2005; McAdam et al., 2007; Jose, 2009; Nair et al., 2009; Giraldo et al., 2011) (see Table 2).

## 5 Animals in agroforestry systems

Much of an animal's daily behaviour pattern is directed towards attaining or retaining homeostasis so, for example, hunger initiates the seeking and ingesting of food whilst a change in temperature can induce the seeking of shade or shelter. The availability of trees and shrubs can play an important role in the maintenance of homeostasis including providing food, shelter, protection and health benefits (e.g. Fisher, 2007). The behaviour of domestic animals can be largely separated into six categories, namely locomotion, maternal, nutritional, reproductive, resting and social behaviour (Phillips, 1993). Access to trees and shrubs can offer benefits to all behaviour categories except perhaps locomotion, though even this can be improved where trees help with surface drainage.

## 5.1 Social relations and body maintenance

Social relations in animal groups improve in silvopastoral systems when compared to open pasture. Social licking in cattle is directed towards maintaining herd stability and dominant cows actively manage herd stability and their own status within the herd in this way (Šárová et al., 2016). In cattle, the body regions licked by herd mates tend to be those not easily reached through auto-grooming, bringing body maintenance benefits to the receiver (Albright and Arave, 1997). The success of maintaining social order through licking is further reinforced by the licking having a calming effect on the recipient animal by lowering the heart rate (Laister et al., 2011). Cattle stay closer together in silvopasture and Améndola et al. (2013) reported that social licking constituted 78% of all social interactions compared to only 41% on open pasture with few or no trees. The beneficial effect of trees on social cohesion is also partially attributed to the provision of shade from solar radiation and to reduced visibility of herd members (Mancera and Galindo, 2011; Schütz et al., 2010; Chamove and Grimmer, 1993). Human-animal relationships can also benefit from silvopasture since, where there is access to trees and shrubs behind which partial concealment is possible, good human-animal relations can be promoted, leading to a reduction in anxiety in the animals and improved handling (Ocampo et al., 2011).

Grooming is an innate behaviour that aids the removal of dead skin and hair from the coat. Farm animals use their environment, particularly trees, to help maintain coat condition (see Fig. 5) and offering them good access to a variety of heights and angles such as tree trunks and low-slung branches allows them to access most body parts. In this way, they can effectively dislodge

**Figure 5** Cow and sheep rubbing on low-hanging branches.

external parasites such as ticks and, similarly, plant seeds that can penetrate the skin can also be removed (Mooring and Samuel, 1998). At moulting time, access to sufficient scratching posts becomes particularly important and increasingly so for those sheep breeds now being selected or developed for fleece shedding. Sheep rubbing themselves can also alert a stockperson to fly strike or mite problems. Although hens dust bathe and use their beaks to preen their feathers, body maintenance is promoted in the presence of trees since preening levels are higher when hens are sheltered by tree canopies than when on open ground (Larsen et al., 2017).

## 5.2 Shade

Globally, the most important role of trees for domestic animals is to provide shade from the sun and Karki and Goodman (2009) recorded up to 58% less solar radiation on silvopasture compared to open pasture. Additional cooling occurs from improved air circulation around trees and through leaf moisture evaporation, particularly with broadleaf trees (Shashua-Bar et al., 2009). Cattle begin seeking shelter at about 21°C, depending on acclimatisation, breed, coat colour and yield status. Seeking shade is an effective strategy since the skin temperature of cattle on silvopasture has been recorded at 4°C lower than that of cattle on open pasture (Betteridge et al., 2012).

In hot weather, a lack of shade increases respiration rate and animals become less engaged with normal activities (Jose et al., 2004). The stress of coping with high temperatures reduces feeding behaviour and increases loafing behaviour. No access to shade reduced milk production in dairy cows (Fisher et al., 2008) and Mitlohner et al. (2001) calculated that beef cattle with no shade took almost three weeks longer to reach a target weight than animals with shade. In contrast, animals in silvopastoral systems engage in resting bouts of better quality, graze more than loaf and the landscape is utilised more evenly than open pasture with few or no trees (Pent, 2017; Améndola et al., 2013; Karki and Goodman, 2009). Silvopasture and its extended canopy cover therefore not only offers animals a way to manage solar heat by enabling them to carry out natural shade-seeking behaviour, it encourages the maintenance of feeding behaviour which in turn maintains growth and production rates in a way that is very much in keeping with sustainable, organic farming practice (Fig. 6).

The cooling effect of trees can further promote health since analysis of weather conditions and seven years of bulk milk data from 3727 dairy farms in northern Italy showed a positive correlation between temperature–humidity index and somatic cell counts. These data also showed a negative correlation with milk fat and protein percentages (Bertocchi et al., 2014), with a similar response of reduced fat and protein content in milk being recorded in New Zealand (Bryant et al., 2007) indicating a reduced milk quality from cows suffering heat stress.

**Figure 6** Goats and dairy cows seeking shade under trees.

## 5.3 Shelter from wind, cold and rain

Warmer winds can be beneficial for livestock welfare, particularly where flies and midges are a problem since even the relatively large blow fly has poor body control in winds of 9 kph and greater. Where sheep are at risk of fly strike, moving them to pasture with a higher wind speed is a recommended practice. Cold winds, on the other hand, can lower the effective temperature considerably below air temperature. For example, with an air temperature of 2°C and a wind speed of 24 kph, the effective temperature becomes −7°C. The thermoneutral zone defines the upper and lower critical temperatures at which environmental conditions become detrimental to the health and well-being of an animal (Silanikove, 2000). For beef cattle management, the general rule for feeding is that for every 1°C drop in temperature below the lower critical temperature, there is a 2% increase in energy requirements. However, merely feeding more without offering shelter would be an insufficient care response when considering organic animal welfare since the seeking of shelter is a most basic behaviour for all animals and a prerequisite for good welfare. As Gregory (1997) noted, *'providing shelter is a moral responsibility'* and appropriate shelter has been specifically named in the second of the Five Freedoms since their publication in 1979 (FAWC, 1979).

Tree canopies and shelter belts offer animals protection from wind, rain and cold. The presence of trees can also act as a buffer against temperature fluctuations in cold weather when minimum grass temperatures can be raised by 6°C (Percival et al., 1984a). A windbreak offers the most effective shelter from driving wind and can consist of high hedges, one or more rows of trees or trees and shrubs combined. Understanding how shelter belts function is important since several characteristics have an influence on the resulting levels of shelter (e.g. Tamang et al., 2009). For example, porosity affects turbulence levels on the leeward side and shelterbelts that allow some wind through create less disturbance than denser growth. A porosity level between 40% and 60% is recommended, although open bottoms should be avoided to stop the acceleration of wind at ground level (Gregory, 1997). Depending on both tree species and the density of the shelter belt, the lowest wind speed occurs between 2 and 6 times the distance of windbreak height. Of course, the direction of the prevailing wind will also determine how effective windbreaks will be with the most effective being perpendicular to the wind or at a right angle if wind is a problem from more than one direction.

Trees with an alternative primary function can provide animals with good shelter. Pigs, for example, can hide and forage in biofuel plantations and areas of densely planted pines, designated for harvesting, can offer animals a living barn during both summer and winter (e.g. Orefice, 2015). They may also offer additional protection from insects since pine species have insect repellent properties (Maia and Moore, 2011). For broadleaf trees, there is anecdotal evidence that some deciduous tree species are less attractive to flies and these include elderberry, lime and walnut. Elder trees were traditionally planted outside pantry windows for this reason.

## 5.4 Shelter for newborn offspring

Whether animals are hider or follower species at the beginning of life, trees and shrubs can play a role in their survival and protection. With full access to natural resources, the use of trees features heavily in the nest-building of sows. First, a pit is dug and lined with grasses and leaves and this is then fully covered with larger branches providing protection to the piglets from both weather and predators. Cattle and deer are hider species and utilise trees, shrubs and other vegetation to hide their newborn offspring (Lidfors et al., 1994). These neonates can remain hidden for up to several days, with dams visiting them at intervals to feed and tend them. Under farmed conditions, a lack of opportunities to hide can lead to aberrant behaviour in both red deer hinds and calves (Janiszewski et al., 2016; Wass et al., 2004).

Although sheep are a follower species, with lambs able to stand and walk soon after birth, adequate shelter can play an important role in their survival. Cold-induced peak metabolic rate is the maximum rate of resting metabolic thermogenesis without any heat production associated with exercise (Minnaar et al., 2014; Wiersma et al., 2007). For resting juvenile animals, suffering from cold stress, peak metabolic rate is unsustainable since they have limited body reserves. Lambs born outdoors can lose body heat very quickly in the first hours of life and as much as 10°C in the first 30 min (McCutcheon et al., 1981). Exposure and starvation combined causes about 30% of all lamb deaths (McCutcheon et al., 1981; Kerslake et al., 2005). They are therefore reliant upon the external environment for shelter from low temperatures, wind, rain or snow (see Fig. 7).

**Figure 7** With shelter, ewes stay for longer at the birth site, promoting the mother-offspring bond.

After birth, ewes form a strong bond with their lambs from a period of 20-30 min of grooming and there is a 5-6 h window in which this optimum bonding can occur. Where shelter is offered in proximity to food and water sources, ewes spend longer at the birth site, increasing the likelihood of a strong ewe-lamb bond being formed (Alexander et al., 1984). Unsurprisingly, the survival and well-being of youngstock increases with access to shelter in inclement weather. Lambs offered shelter during the first three weeks of life have a higher growth rate than those without shelter (Alexander and Lynch, 1976). The mother-offspring bond is fully complete after 1 day for singles and 3 days for twins and the lamb and ewe can recognise each other visually and vocally. At this point, lambs spend more time in sheltered areas than the ewes, developing their own behaviour patterns appropriate for their needs in maintaining homeostasis. In extensive rough-terrain landscapes, a secondary benefit of providing trees for shade and shelter is that, since sheep congregate there, they become easier to shepherd.

## 6 Trees as a source of nutrition and medicine

### 6.1 Browsing behaviour

When sourcing food, herbivores divide their feeding time between ingesting known food sources and exploring for new or better sources (Huzzey et al., 2013). The silvopastoral system delivers a variety of feed sources and particularly in the three-level pastoral system which includes trees, herb-rich pasture and shrubs grown specifically for browsing. Hedges, or living fences, also offer good browsing opportunities as well as providing habitat for native flora and fauna (Pulido-Santacruz and Renjifo, 2011).

Cattle are classified as grazers, sheep as non-selective with a preference for grazing and goats as intermediate selective animals with a preference for browsing (Dicko and Sikena, 1992). Their preferred diet is partly reflected in mouth physiology where the cleft upper lip of goats and sheep allows for greater manipulation of individual plant items than does the muzzle of cattle when, for example, selecting leaves and avoiding thorns (Shipley, 1999). For cattle, the average intake of browse is 12% of total diet, for sheep this is 20% and for goats, it is 60%. Intake of browse not only varies widely between animal species, it also varies with available tree species and season (Wilson, 1969). In a temperate climate, intake increases with access to young spring growth (Vandermeulen et al., 2016), or when the ground is covered in snow. In regions with a hot, dry season, when grasses and forbs do not thrive, intake of browse increases substantially (Solorio Sanchez and Solorio Sanchez, 2002). Then, browse intake can increase to 55% for cattle, 76% for sheep and 93% for goats (Dicko and Sikena, 1992).

Although the gastrointestinal tract of cattle, with its large rumen, is well adapted to grass diets it does not inhibit the efficient digestion of browse (Clauss and Hofmann, 2014). Observations of cattle, in temperate conditions, given access to fresh pasture show initially a high level of interest in browse opportunities upon entry to the field. When different feeds (i.e. grasses, forbs and browse) are readily available, the general pattern of cattle is for them to feed predominantly on grazed plants in the morning and evening and then to be more selective in the middle of the day when they ingest more browse (Gupta et al., 1999). Of course, access to browse is not only influenced by availability but also animal agility and motivation. The browse height of sheep is 1.3 m and for cattle it is 2 m (Kemp et al., 2003). In contrast, goats are vertical browsers, browsing on one tree before they move to the next (Hart, 2013). Their physical agility enables them not only to stand on their hind legs to reach forage from the ground, they can also use their front feet to pull down branches as well as climb trees to gain access to higher browse. Goats are well adapted to eating tannin-rich browse and produce a specific saliva capable of binding tannins. Relative to body weight, they also have larger livers than cattle or sheep so they can more effectively process high levels of tannins (Shipley, 1999). Further to the environmental benefits listed in Table 2, feeding browse and other plants with condensed tannins reduces methane emissions and, when total greenhouse emissions (plant, soil and livestock) are considered, open pasture tends to be net emitters whilst well-functioning silvopasture tends to be net sequesterers (Montagnini et al., 2013).

## 6.2 Palatability and feed preferences

Feed preferences are highly influenced by the novelty of a feed source. The strongest influence on feed preference is the dam, first as an unborn foetus and then through the taste in milk. For example, lambs will voluntarily ingest onions and garlic after their mothers had been fed them while pregnant. Watching and copying the dam continue the strong maternal influence on

**Figure 8** Hereford cattle browsing on a mixed-species hedgerow.

food preferences (Provenza et al., 2015; Thorhallsdottir et al., 1990). The next strongest influence is peer behaviour (see Fig. 8) and for animals faced with novelty, copying peer behaviour helps to create shortcuts in learning, reducing both the risks associated with adopting novel behaviours and increasing the speed of adoption. The third influence on feed preferences is time, with increasing familiarity overcoming neophobia. In trials feeding willow species as a novel feed source, cattle intake increased from 1.5 to 3.5 kg after 81 days with access and the diameter of wood ingested also increased from 4 mm to 8 mm (Moore et al., 2003). Similarly, with lambs, the diameter of wood ingested increased from 3 mm to 4.2 mm after a 10-week period (Diaz Lira et al., 2008). Additionally, familiarity with one browse species encourages the browsing of other species (Burritt and Frost, 2006).

The palatability of available trees was ranked for the Woodland Grazing Toolbox (Forestry Commission, Scotland, 2016). Of the species included in the ranking, aspen and willow were ranked highest followed by ash and rowan, hazel and oak, scots pine, juniper and holly, birch and hawthorn, beech and finally alder as the least palatable. Note that hawthorn is ranked relatively low in palatability but the toolbox acknowledges that this may be too low, given the high level of browsing observed. Determining palatability is not straightforward since this is not simply defined by the taste of something. Instead, it is the interrelationship between the senses (primarily taste) and feedback from post-ingestive processes so that both the chemical characteristics of food and an animal's physiological state can influence what and how much of a food source is eaten (Kearney et al., 2016).

Animals that have a nutrient deficiency have demonstrated an ability to seek out sources of that nutrient when there are defining features that reveal its presence such as a particular smell or taste (Villalba et al., 2008; Villalba and Provenza, 2009). For example, cattle managed on moorland can be deficient in phosphorus and calcium and, without supplementation, seek alternative even abnormal sources such as animal bones to redress the balance. Lambs born in regions with a dry season, when there is little fresh green feed available, can be susceptible to white muscle disease from the lack of selenium and/or vitamin E. Drought-tolerant browse can offer a good source of vitamin E during this period but palatability is low from the high level of bitter plant defence chemicals. Nevertheless, Amanoel et al. (2016) demonstrated the willingness of lambs, deficient in vitamin E, to preferentially select a less palatable feed with a high content of vitamin E. The authors suggest that post-ingestive influences on intake can occur beyond the gut level, reasoning that the metabolic effects of the antioxidant properties of vitamin E take place at the tissue level. Additionally, they noted that internal feedback mechanisms are sufficiently sensitive for animals to seek out a nutrient supply before the emergence of any clinical symptoms associated with a deficiency. Sensitivity to nutrient deficiency and an ability to identify and seek out feed sources with specific nutrients indicate that animals may well be able to monitor and manage their own nutritional health to a much greater degree in a sufficiently diverse environment, reinforcing the importance of diversity to the Principle of Health and resilience in organic systems.

## 6.3 Nutrition and health

In general, browse is a good source of nutrition for farm animals comparing favourably with grasses grown in the same environment. Emile et al. (2016) analysed crude and degradable protein in the leaves of different tree species as well as alfalfa/lucerne and ryegrass, finding that several tree species, particularly ash, lime and mulberry, compare very favourably with the grasses. When animals are offered a varied diet they tend to eat more of everything, including the less palatable feedstuffs with higher palatable feed acting as a buffer. Additionally, shrubs and trees, in combination with pasture plants, produce more forage per unit area of land than pasture plants alone. Tree species that are relatively low in nutritional value can offer bulk for animals managed on restricted feed. Dry sows are typically fed on restricted rations to prevent obesity, leading to hunger, frustration and the risk of abnormal behaviour such as stone chewing. Offering vegetation, including browse, reduces the hunger problem as well as frustration by enabling sows to engage in natural foraging behaviour; an approach that is utilised in Brazil (Arey and Brooke, 2016).

Browse is a very good source of micronutrients and particularly minerals (Luske and van Eekeren, 2014). In silvopasture, the nutrients retrieved by the deeper-rooted trees become available to the animal through direct browsing.

Furthermore, mineral content is higher in dried tree fodder than fresh, increasing its value as a sustainable source of minerals (Smith et al., 2017). Zinc, for example, is present in all animal tissue, organs and bones and plays an important role in many biological functions including the immune system and skin integrity. Additionally, zinc influences the utilisation of nutrients and a deficiency can impair both protein and carbohydrate metabolism (Blair, 2011). Of the different tree species, zinc is particularly high in willow, with *Salix viminalis* containing 245 mg/kg DM (Robinson et al., 2005), and substantially higher than in grass at 5 mg/kg DM, silage at 25-30 mg/kg DM or hay at 17-21 mg/kg DM (McDonald et al., 1995). In New Zealand, the fungus *Pithomyces chartarum* is prevalent in pasture, and the spores cause facial eczema in cattle. Zinc supplements can prevent the eczema but browsing on willow has been shown to be more effective than drenching (Anderson et al., 2012a,b).

## 6.4 Plant secondary metabolites

As a defence against attack from threats including pathogens, insects and browsing herbivores, plants produce a range of chemicals collectively known as plant secondary metabolites (PSMs). Bitter tasting tannins combat defoliation from insects and herbivores by, for example, reducing palatability whereas salicylic acid aids plants in their defence against pathogens (Taiz and Zeiger, 2010). Condensed tannins not only inhibit intake through their bitter taste, they also reduce the digestibility of protein in the rumen by binding them to enzymes. However, these enzymes are themselves broken down in the more acidic abomasum effectively delivering higher-quality rumen-bypass protein to the small intestine. For a review of positive and negative effects of dietary tannin on ruminal fermentation, see Addisu (2016). Historically considered to adversely affect herbivores, with little nutritional benefit, the presence of tannins is now recognised for its role in providing a valuable source of rumen-bypass protein (Min et al., 2003). Under arid farming conditions, a diet including browse with condensed tannins improves wool quality both in terms of weight and staple length compared to the wool from sheep not offered browse (Ramírez-Restrepo et al., 2005). Additionally, there is an increase in the fertility of ewes with more lambs being born and surviving to weaning age compared to ewes only offered drought pasture (Pitta et al., 2004).

## 6.5 Parasite control

Animals are not only able to identify sources of specific nutrients, they can also utilise plants to help manage health needs such as parasite burdens. Taste and nutritional value diminish in favour of other chemical components such as condensed tannins that offer some parasite control. Access to browse can support an animal's resilience to gastrointestinal (GI) parasites in several ways. The increase in available protein not only improves general health and production levels, it can also reduce susceptibility to GI parasites (Walkden-Brown and Kahn, 2002) as well as playing a potential role in future immunity to parasite infection. The tannins themselves have a direct anthelmintic effect on several parasites of the intestinal tract in grazing animals, including *Ascaris* spp. limiting the number of eggs hatching, the number of larvae maturing to adults as well as the size of those that do mature so that these smaller adults produce fewer eggs (Williams et al., 2014; Novobilský et al., 2011; Waller et al., 2001). Studies of sheep and goats feeding on tannin-rich browse show up to 50% reductions in faecal egg counts (Min and Hart, 2003). The mineral copper has a similar effect as condensed tannins on GI parasites, particularly *Haemonchus contortus* and the tree species hazel and beech are both good sources of copper (Luske and van Eekeren, 2015). Where sheep have developed resistance to GI parasites in the form of a parasite-specific immunoglobulin (IgA), goats have developed resilience through feeding habits that physically remove them from the risk zone which is up to approximately 15 cm from soil level. Along with the tannins contained in browse species, the elevated height of browse can be used as part of a parasite management programme for species other than goats at risk of heavy parasite burdens. Feeding above the risk zone has the additional benefit of interrupting the life cycle of some GI parasites (Halvorson et al., 2011; Hart, 2013). Importantly, if parasitised herbivores learn to self-treat, given a variety of plants with a variety of PSMs, producers need not resort to giving fixed doses of chemicals to all animals in the herd, likely with different parasite burdens. This not only maximises the use and self-sufficiency of natural resources it also reduces reliance on veterinary interventions (Sundrum, 2001).

## 6.6 Pain management

Evidence of self-medication in animals, or zoopharmacognosy, is widespread though often anecdotal. There are however some studies on apes which show convincing evidence of plants being eaten to cause vomiting, to relieve constipation or to manage intestinal parasite burdens (Shurkin, 2014). There is also increasing evidence that farm animals are capable of making associations between food containing medication and treatment of disease (Villalba and Provenza, 2007). For example, broiler chickens grow rapidly and can suffer from leg problems as a result. When offered normal feed or their normal feed with added pain medicine, lame broilers not only consistently selected the treated feed more than sound birds, they increased their intake in line with the severity of lameness (Danbury et al., 2000). This indicates that not only do lame boilers suffer pain but that they are indeed capable of seeking pain relief through feedstuffs. In a different study, a mixed group of sheep were fed illness-inducing levels of either grains, tannins or oxalic acid. The sheep were then offered access to three medicines (sodium bentonite, polyethylene glycol (PEG) and di-calcium phosphate) only one of which would aid recovery. The sheep learned to select the appropriate medicine to restore health (Villalba et al., 2006). The presence of mothers can also influence the self-medication behaviour of offspring. Lambs with mothers ingested higher levels of PEG after feeding on tannin-rich feed than did lambs with no mother present. Prior experience of the ewe to the medicinal effects of PEG appeared to be less important than their physical presence on lamb behaviour (Sanga et al., 2011).

The PSM, salicylic acid, is a recognised pain suppressant and is the origin of the pain killer aspirin. It also has anti-inflammatory and mild antibiotic properties and although it is present in a wide range of plants, it is abundant in some tree species such as willow and poplar. When offered a choice, sheep choose to browse on willow and poplar with high salicylic acid content, possibly due to the beneficial anti-inflammatory effects. Learning to self-medicate, or as Provenza (1995) writes acquiring 'nutritional wisdom', requires a landscape rich in diversity and for animals to have sufficient access to enable associations to be made (Provenza et al., 2015).

## 7 Challenges in integrated livestock and forestry systems

### 7.1 Development and management of integrated systems

Developing and managing integrated systems requires sound understanding of both the system and of the local geographical, environmental and animal needs. A significant problem in making a transition to ICLS is the initial cost of establishment (Franzluebbers, 2007) and the establishment itself requires landownership or sound long-term tenancy agreements. A further cost can be the acquisition of animals for fattening (Allen et al., 2007; Martha et al., 2011), and Hendrickson et al. (2008) have also highlighted the need for additional labour in some systems. The implementation of diversified and integrated systems is not always translated into benefits, particularly when crop, livestock and grasslands are not fully integrated. In these circumstances, negative interactions may occur between the components of the system due to competition for resources (Jose et al., 2004; Gea-Izquierdo et al., 2009; Gea-Izquierdo and Cañellas, 2009), to the detriment of animal welfare. It has been noted, for example, that, in some smallholder ICLS in developing countries, the pressure on land for crops as well as the value of manure has led to smallholders keeping animals indoors rather than allowing them to graze and browse freely outside (Muwanga et al., 2011; Edwards et al., 2010). Quality of management is key to success (Bonaudo et al., 2014).

Once established, agroforestry systems require disciplined and skilled management of both plants and animals (e.g. Le Houérou, 2000). In well-designed and well-managed silvopastoral systems, the overall herbage yield increases when compared to open pasture, however, understanding how to manage the light requirements of different tree and grass species is important (Orefice, 2015). High tree densities and high canopy density can negatively affect the growth of some species of grass in temperate environments with good rainfall but improve yields in dryland farming conditions (Esquivel-Mimenza et al., 2013; Hemery et al., 2005; Devkota et al., 2009). Rozados-Lorenzo et al. (2007) found pasture growth to be stable under young trees planted at 556 trees per ha but under higher tree densities, pasture production decreased markedly. As trees are slow growing compared to other plant resources, initial investments can be tied for several years before showing returns. Browsing is not recommended until trees are three years old and they also require protection from wild herbivores, at least in the first few years of growth. Similarly, biodiversity can be higher in well-designed and well-managed silvopastoral systems that include grazing livestock but species richness is threatened where a high stocking density creates high grazing pressure (Robertson et al., 2012a,b; Seffan-Dewenter and Leschke, 2003; Bergmeier et al., 2010). Under these conditions, animals are also more likely to damage trees by debarking them and the growth rates of juvenile animals can be checked (SSBA, 2008). Conversely, loss of habitat can also be caused by the under-grazing of traditional woodland pasture systems (Bianchetto et al., 2015).

High levels of knowledge and technical skills are needed to understand the trade-offs occurring within ICLS. Moraine et al. (2014) and Ryschawy et al. (2015) have highlighted the importance of knowledge exchange and cooperation among farmers. The variability in crop-livestock farms also make it necessary to increase interdisciplinary research to support farmers (Gibon et al., 2010). Moreover, this research must be carried out by working in conjunction with farmers, agricultural advisers, policy decision-makers and other rural stakeholders (Tarawali et al., 2011; Gibon et al., 2012). Government training and support is an important factor in maintaining local mixed crop-livestock farms in developed countries in Europe (Ryschawy et al., 2014b). Conservation agriculture practices have been highlighted as a way to improve ICLS (Marchão et al., 2009; Escribano et al., 2014, 2015; Escribano, 2016a,b). Ryschawy et al. (2015) have recommended diversifying rotations while integrating grasslands into the system. Sowing forage legumes between two cash crops to achieve herd feed self-sufficiency, while maintaining soil fertility, is one approach (Ryschawy et al., 2014a,b). Crop residues, such as maize residues, can be an excellent source of biomass for ruminant feed. In general, such studies have found that beef cows are able to valorise both forage and crop residues, whilst calves can be fed grain during preconditioning and finishing (Russelle et al., 2007).

## 7.2 Aspects of tree and animal interactions

Although tree fodder can be fed as direct browse, fresh-cut fodder or preserved fodder (dried or ensiled), trees cannot necessarily offer year-round provision of food or medicine and managing grazing in the presence of dormant browse species presents a management problem, particularly for highly palatable species (Poudel et al., 2017). Managing levels of browse during the growing season can be difficult without some control of access, particularly with the regeneration of existing shrub species and the establishment of young trees. Managing stocking rates to avoid more than 50% defoliation from browsing is recommended since heavier browsing can reduce a plant's ability to regrow (Hendrickson and Olson, 2006). How browse is managed will depend partly on its role in feed management and partly on species. Sustainable browsing, where browse is defoliated at 50% or less, requires a minimum of 8 weeks rest (Hart, 2013). Understanding the relative palatability of different trees and grazed plants is also important in assessing defoliation rates.

Whilst poultry thrive in established silvopastoral systems there appear to be limited health or welfare benefits from newly planted woodland, at least during the first two years of establishment and growth (Jones et al., 2007). That said, younger trees may still offer some protection against avian influenza. Wild birds can spread avian influenza to poultry by using their home range but the presence of sufficient trees on pasture appears to play an important role in protecting them. Wild birds can be grouped into three categories depending on the risk they pose to domestic farm birds and the high-risk wild birds, including waterfowl, are less likely to congregate in areas with more than 5% cover from trees (Bestman et al., 2017). Geography also plays a role in how the presence of trees can impact animal handling. Whilst shepherding can be improved in upland terrain, both stockperson overview and the gathering of stock can take longer on lowland silvopasture including orchard grazing systems compared to open pasture (Percival, 1984b; Burgess et al., 2017).

Physical, as well as chemical, plant defences can be a problem for browsing animals. Thorns on trees are less of an issue early in the growing season when they are still young and soft but, once hardened, the thorns can cause real injuries. For example, blackthorn puncture wounds can become swollen and infected. Physical plant defences also slow down the speed of feeding, reducing both bite rate and the size of bites. Once the tips have been eaten, the tree can form a mat of growth difficult to penetrate; however, this can provide an ideal protective environment in which a young tree sapling can grow. Chemical defences not only affect the efficiency of digestion, they can also change the order in which plants are eaten by herbivores or reduce intake through low palatability (Papachristou et al., 2005). Although the presence of tannins in feed can be beneficial at around 1–4% of dry matter intake, an increase beyond 5% can cause digestibility problems (Addisu, 2016). Hydrolysable tannins are potentially poisonous, although most ruminants are able to adjust to a diet containing them (Waghorn and McNabb, 2003). Bhat et al. (2013) stated that animals are capable of self-regulating intake of different plants but this is only possible when there is sufficient diversity in forage sources for them to avoid plants with high levels of PSMs.

Where too little shade is offered for the number of animals present, overcrowding occurs leading to an increase in risk of disease, parasite contamination, overheating, death of vegetation and surface soil compaction (e.g. Bray and Lancaster, 1992). Too little shelter at lambing time and subsequent overcrowding reduces the survival of lambs from starvation, exposure and mis-mothering (Robertson et al., 2012a,b). Negative impacts on plant and animal communities other than the farmed species may also occur depending on the placement of trees within the landscape by, for example, encouraging animals seeking shade to congregate along waterways with ecologically fragile banks (Jose et al., 2004).

## 7.3 Trees and flies

Trees, particularly those on the edges of woodland, can attract flies such as blow flies, nostril flies and head flies. Flies can cause distress and injury and they are implicated in the spread of diseases such as pinkeye (or infectious bovine keratoconjunctivitis) and summer mastitis (Radostits et al., 2007; Hillerton et al., 1983). Animals show a range of behaviours to try to avoid or cope with flies. These can be passive behaviours such as hiding in thickets or long grass, lying passively, trying to hide body parts most at risk and crowding together. More active behaviours include fleeing from the presence of biting flies or frequent moving to different areas of their environment. An avoidance of any shade in hot weather can indicate that flies are present, although use of shade can also be affected by distance to water. Oliveira et al. (2013) found that cows did not use offered shade when the distance to water was 40 m but they spent 26% of the daytime in shade when the trough was placed only 5 m away.

Fly burdens and associated diseases can, nevertheless, be better managed with a greater understanding of how they interact with trees and of local environmental conditions. Simplifying landscapes through the reduction of trees as well as non-crop areas reduces biodiversity and natural pest control functions (Bianchi et al., 2006). In contrast, in established silvopastoral systems, there are more species of dung beetles present and each beetle is more active than those on open pasture, ensuring that dung is removed more quickly from the surface (Menegaz De Farias et al., 2015; Nair and Graetz, 2004). In more diverse and structurally complex systems, there are also a significant number of predators of small insects, including birds (Gillespie et al., 1995). Predator-prey interactions that lead to pest control in agroforestry systems can be explained by the theory of 'three trophic level terrestrial interactions' whereby the introduction of a predator to a system will lead to a reduction in prey density (Price et al., 1980). Under these conditions, the number of head flies trapped on silvopasture was 40% lower than those trapped on open pasture. For diverse farming systems, chickens can also be considered a useful predator of insects.

## 8 Conclusion

The sustainability of animal welfare offered by well-designed and well-managed agroforestry systems is higher compared to other livestock systems (Broom, 2017). Diversity offers opportunities for learning, enabling animals to respond appropriately to internal and external factors affecting homeostasis. Choice of food and environmental diversity enables animals to exert a level of control so their behaviour patterns more closely reflect what can be considered natural, or normal. Social behaviour is also more positive than on open pasture. As a farming system, agroforestry improves social harmony and offers diversity in feed and environment and can therefore be considered to offer good welfare conditions that are a prerequisite for organic farming, in accordance with the principles of health, ecology, fairness and care (Améndola et al., 2016; IFOAM, 2005). As Sommerville and Jones (2013) stated, *'Achieving a high quality of life for farm animals requires provision rather than deprivation'.*

## 9 Where to look for further information

AGFORWARD (2014–17). AgroFORestry that Will Advance Rural Development. EU's Seventh Framework Programme for Research and Technological Development. https://www.agforward.eu/index.php/en/

AFINET (2017–20). Agroforestry Innovation Networks and Regional Agroforestry Innovation Networks (RAINS). http://www.eurafagroforestry.eu/afinet

Vandermeulen, S., Ramírez-Restrepo, C. A., Beckers, Y., Claessens, H. and Bindelle, J. (2018). Agroforestry for ruminants: a review of trees and shrubs as fodder in silvopastoral temperate and tropical production systems. *Animal Production Science*, 58, pp. 767–77.

## 10 References

Acosta-Martínez, V., Zobeck, T. M. and Allen, V. (2004). Soil microbial, chemical and physical properties in continuous cotton and integrated crop-livestock systems. *Soil Science Society of America Journal*, 68(6), pp. 1875–84.

Addisu, S. (2016). Effect of dietary tannin source feeds on ruminal fermentation and production of cattle: A review. *Online Journal of Animal and Feed Research*, 6, pp. 46–56.

Albright, J. L. and Arave, C. W. (1997). *The Behaviour of Cattle*. CABI Publishing, Wallingford UK.

Alexander, G. and Lynch, J. J. (1976). Phalaris windbreaks for shorn and fleeced lambing ewes. *Proceedings of the Australian Society of Animal Production*, 11, pp. 161–4.

Alexander, G., Kilgour, R., Stevens, D. and Bradley, L. R. (1984). The effect of experience on twin-care in New Zealand Romney sheep. *Applied Animal Behaviour Science*, 12, pp. 363–72.

Allen, V. G., Brown, C. P., Kellison, R., Segarra, E., Wheeler, T., Dotray, P. A., Conkwright, J.C., Green, C. J. and Acosta-Martinez, V. (2005). Integrating cotton and beef production to reduce water withdrawal from the Ogallala Aquifer in the Southern High Plains. *Agronomy Journal*, 97, pp. 556–67.

Allen, V. G., Baker, M. T., Segarra, E. and Brown, C. P. (2007). Integrated irrigated crop-livestock systems in dry climates. *Agronomy Journal*, 99, pp. 346–60.

Amanoel, D. E., Thomas, D. T., Blache, D., Milton, J. T. B., Wilmot, M. G., Revell, D. K. and Norman, H. C. (2016). Sheep deficient in vitamin E preferentially select for a feed with a higher concentration of vitamin E. *Animal*, 10, pp. 183–91.

Améndola, L., Solorio, F. J., González-Rebeles, C. and Galindo, F. (2013). Behavioural indicators of cattle welfare in silvopastoral systems in the tropics of México. In: M. J. Hötzel and L. C. P. M. Filho (Eds), *Proceedings of 47th Congress of International Society for Applied Ethology*, 2–6 June 2013, Florianópolis, Brazil, p.150.

Améndola, L., Solorio, F. J., Ku-Vera, J. C., Améndola-Massiotti, R. D., Zarza, H. and Galindo, F. (2016). Social behaviour of cattle in tropical silvopastoral and monoculture systems. *Animal*, 10, pp. 863–7.

Anderson, V. and Schatz, B. (2003). Biological and economic synergies and methods of integrated beef cow and field crops. Unified Beef Cattle and Range Research Report, North Dakota State University, Fargo, ND.

Anderson, C. W. N., Robinson, B. H., West D. M., Clucas, L. and Portmann, D. (2012a). Zinc-enriched and zinc-biofortified feed as a possible animal remedy in pastoral agriculture: Animal health and environmental benefits. *Journal of Geochemical Exploration*, 121, pp. 30–5.

Anderson, D. M., Fredrickson, E. L. and Estell, R. E. (2012b). Managing livestock using animal behavior: Mixed-species stocking and flerds. *Animal*, 6(8), pp. 1339–49.

Arey, D. and Brooke, P. (2016). *Animal Welfare Aspects of Good Agricultural Practice: Pig Production*. Compassion in World Farming, Surrey, UK, p. 166.

Behera, U. K., Kaechele, H. and France, J. (2015). Integrated animal and cropping systems in single and multi-objective frameworks for enhancing the livelihood security of farmers and agricultural sustainability in Northern India. *Animal Production Science*, 55(10), pp. 1338–46.

Benavides, R., Douglas, G. B. and Osoro, K. (2009). Silvopastoralism in New Zealand: Review of effects of evergreen and deciduous trees on pasture dynamics. *Agroforestry Systems*, 76, pp. 327–50.

Bene, J. G., Beall, H. W. and Côté, A. (1977). *Trees, Food and People: Land Management in the Tropics*. IDRC, Ottawa, Canada.

Bergmeier, E., Petermann, J. and Schröder, E. (2010). Geobotanical survey of wood-pasture habitats in Europe: Diversity, threats and conservation. *Biodiversity and Conservation*, 19, pp. 2995–3014.

Bertocchi, L., Vitali, A., Lacetera, N., Nardone, A., Varisco, G. and Bernabucci, U. (2014). Seasonal variations in the composition of Holstein cow's milk and temperature-humidity index relationship. *Animal*, 8, pp. 667-74.

Bestman, M., de Jong, W., Wagenaar, J.-P. and Weerts, T. (2017). Presence of avian influenza risk birds in and around poultry free-range areas in relation to range vegetation and openness of surrounding landscape. *Agroforestry Systems*. DOI:10.1007/s10457-017-0117-2.

Betteridge, K., Costall, D., Martin, S., Reidy, B., Stead, A. and Millner, I. (2012). Impact of shade trees on Angus cow behaviour and physiology in summer dry hill country: Grazing activity, skin temperature and nutrient transfer issues. In: *Advanced Nutrient Management: Gains from the Past - Goals for the Future*. Occasional Report No. 25. Fertilizer and Lime Research Centre, Massey University, Palmerston North, New Zealand.

Bhat, T. K., Kannan, A., Singh, B. and Sharma O. P. (2013). Value addition of feed and fodder by alleviating the antinutritional effects of tannins. *Agricultural Research*, 2, pp. 189-206.

Bianchetto, E., Buscemi, I., Corona, P., Giardina, G., La Mantia, T. and Pasta, S. (2015). Fitting the stocking rate with pastoral resources to manage and preserve Mediterranean forest lands: A case study. *Sustainability*, 7(6), pp. 7232-44.

Bianchi, F. J. J. A., Booij, C. J. H. and Tscharntke, T. (2006). Sustainable pest regulation in agricultural landscapes: A review on landscape composition, biodiversity and natural pest control. *Proceedings of the Royal Society B*, 273, pp. 1715-27.

Bird, P. R. (1998). Tree windbreaks and shelter benefits to pasture in temperate grazing systems. *Agroforestry Systems*, 41(1), pp. 35-54.

Blair, R. (2011). *Nutrition and Feeding of Organic Cattle*. CABI, Wallingford UK.

Bonaudo, T., Bendahan, A. B., Sabatier, R. and Tichit, M. (2014). Agroecological principles for the redesign of integrated crop-livestock systems. *European Journal of Agronomy*, 57, pp. 43-51.

Borin, M., Passoni, M., Thiene, M. and Tempesta, T. (2009). Multiple functions of buffer strips in farming areas. *European Journal of Agronomy*, 32, pp. 103-11.

Brandle, J. R., Hodges, L. and Zhou, X. (2004). Windbreaks in North American agricultural systems. *Agroforestry Systems*, 61, pp. 65-78.

Bray, T. S. and Lancaster, M. B. (1992). The parasitic status of land used by free range hens. *British Poultry Science*, 33, pp. 1119-20.

Broom, D. M. (2017). Components of sustainable animal production and the use of silvopastoral systems. *Revista Brasileira de Zootecnia*, 46, pp. 683-8.

Bryant, J. R., Lopez-Villalobos, N., Pryce, J. E., Holmes, C. W., Johnson, D. L. and Garrick, D. J. (2007). Environmental sensitivity in New Zealand dairy cattle. *Journal of Dairy Science*, 90, pp. 1538-47.

Burgess, P., Chinery, F., Eriksson, G., Pershagen, E., Pérez-Casenave, C., Upson, M., García de Jalón, S., Giannitsopoulos, M. and Graves, A. (2017). System report: Grazed orchards in England and Wales. Agroforestry for high value tree systems. Contribution to Deliverable 5.14: Lessons learnt from innovations related to agroforestry for livestock for EU. F.P7 Research Project: AGFORWARD 613520. 22 pp. http://www.agforward.eu/index.php/en/agroforestry-with-ruminants-uk.html

Burritt, E. and Frost, R. (2006). Animal behavior: Principles and practices. In: K. Launchbaugh (Ed.), *Targeted Grazing: A Natural Approach to Vegetation*

*Management and Landscape Enhancement*. American Sheep Industry Association and Cottrell Printing, Centennial, CO, pp. 10-21.

Chamove, A. S. and Grimmer, B. (1993). Reduced visibility lowers bull aggression. *Proceedings of the New Zealand Society of Animal Production*, 53, pp. 207-8.

Chu, B., Goyne, K. W., Anderson, S. H., Lin, C.-H. and Udawatta, R. P. (2010). Veterinary antibiotic sorption to agroforestry buffer, grass buffer and cropland soils. *Agroforestry Systems*, 79, pp. 67-80.

Clauss, M. and Hofmann, R. R. (2014). The digestive system of ruminants, and peculiarities of (wild) cattle. In: M. Melletti and J. Burton (Eds), *Ecology, Evolution and Behaviour of Wild Cattle: Implications for Conservation*. Cambridge University Press, Cambridge, UK, pp. 57-62.

Cook, S. L. and Ma, Z. (2014). Carbon sequestration and private rangelands: Insights from Utah landowners and implications for policy development. *Land Use Policy*, 36, pp. 522-32.

Crespo, G. (2008). Importance of the silvopastoral systems to keep and restore soil fertility in tropical regions. *Cuban Journal of Agricultural Science*, 42(4), pp. 331-7.

Cubbage, F., Balmelli, G., Bussoni, A., Noellemeyer, E., Pachas, A. N., Fassola, H., Colcombet, L., Rossner, B., Frey, G., Dube, F., Lopes de Silva, M., Stevenson, H., Hamilton, J. and Hubbard, W. (2012). Comparing silvopastoral systems and prospects in eight regions of the world. *Agroforestry Systems*, 86(3), pp. 303-14.

Danbury, T. C., Weeks, C. A., Chambers, J. P., Waterman-Pearson, A. E. and Kestin, S. C. (2000). Self-selection of the analgesic drug carprofen by lame broiler chickens. *Veterinary Record*, 146, pp. 307-11.

Devkota, N. R., Kemp, P. D., Hodgson, J., Valentine, I. and Jaya, I. K. D. (2009). Relationship between tree canopy height and the production of pasture species in a silvopastoral system based on alder trees. *Agroforestry Systems*, 76, pp. 363-74.

Diaz Lira, C. M., Barry, T. N. and Pomroy, W. E. (2008). Willow (*Salix spp.*) fodder blocks for growth and sustainable management of internal parasites in grazing lambs. *Animal Feed Science and Technology*, 141, pp. 61-81.

Dicko, M. S. and Sikena, L. K. (1992). Feeding behaviour, quantitative and qualitative intake of browse by domestic ruminants. In: A. Speedy and P. L. Pugliese (Eds), *Legume Trees and other Fodder Trees as Protein Sources for Livestock*. FAO Animal Production and Health Paper 102. FAO, Rome, Italy, pp. 129-44.

Dixon, R. K. (1995). Agroforestry systems: Sources or sinks of greenhouse gases? *Agroforestry Systems*, 31(2), pp. 99-116.

Dumont, B., Fortun-Lamothe, L., Jouven, M., Thomas, M. and Tichit, M. (2013). Prospects from agroecology and industrial ecology for animal production in the 21st century. *Animal*, 7(6), pp. 1028-43.

Dumont, B., Gonzales-Garcia, E., Thomas, M., Fortun-Lamothe, L. and Ducrot, C. (2014). Forty research issues for the redesign of animal production in the 21st century. *Animal*, 8(8), pp. 1382-93.

Dumont, B., Andueza, D., Niderkorn, V., Luscher, A., Porqueddu, C. and Picon-Cochard, C. (2015). A meta-analysis of climate change effects on forage quality in grasslands: Specificities of mountain and Mediterranean areas. *Grass and Forage Science*, 70(2), pp. 239-54.

Edwards, S., Egziabher, T. and Araya, H. (2010). *Successes and Challenges in Ecological Agriculture: Experiences from Tigray, Ethiopia*. FAO, Rome, Italy.

Emile, J. C., Delagarde, R., Barre, P. and Novak, S. (2016). Nutritive value and degradability from temperate woody resources for feeding ruminants in summer. *Proceedings of the 3rd European Agroforestry Conference (EURAF)*, Montpellier France, 23-25 May 2016, p. 468.

Escribano, A. J. (2016a). Organic livestock farming: Challenges, perspectives, and strategies to increase its contribution to the agrifood system's sustainability - A Review. In: Konvalina, P. (Ed.), *Organic Farming - A Promising Way of Food Production*. InTech, New York, NY, pp. 229-60.

Escribano, A. J. (2016b). Beef cattle farms' conversion to the organic system. Recommendations for success in the face of future changes in a global context. *Sustainability*, 8(6), p. 572.

Escribano, A. J., Gaspar, P., Mesías, F. J., Pulido, A. F. and Escribano, M. (2014). A sustainability assessment of organic and conventional beef cattle farms in agroforestry systems: The case of the dehesa rangelands. *Información Técnica Económica Agraria (ITEA)*, 110(4), pp. 343-67.

Escribano, A. J., Gaspar, P., Mesías, F. J., Escribano, M. and Pulido, F. (2015). Comparative sustainability assessment of extensive beef cattle farms in a high nature value agroforestry system. In: V. Squires (Ed.), *Rangeland Ecology, Management and Conservation Benefits*. Nova Science Publishers Inc., New York, NY, pp. 65-85.

Esquivel-Mimenza, H., Ibrahim, M., Harvey, C. A., Benjamin, T. and Sinclair, F. L. (2013). Standing herbage biomass under different tree species dispersed in pastures of cattle farms. *Tropical and Subtropical Agroecosystems*, 16, pp. 277-88.

FAWC (1979). The Five Freedoms. Farm Animal Welfare Council Press Statement, 5 December 1979. http://webarchive.nationalarchives.gov.uk/20121010012428/ http://www.fawc.org.uk/pdf/fivefreedoms1979.pdf (Accessed 19 March 2012).

Fisher, M. W. (2007). Shelter and welfare of pastoral animals in New Zealand. *New Zealand Journal of Agricultural Research*, 50, pp. 347-59.

Fisher, A. D., Roberts, N., Bluett, S. J., Verkerk, G. A. and Matthews, L. R. (2008). Effects of shade provision on the behaviour, body temperature and milk production of grazing dairy cows during a New Zealand summer. *New Zealand Journal of Agricultural Research*, 51, pp. 99-105.

Food and Agriculture Organization of the United Nations (FAO) (2001). *Mixed Crop-livestock Farming. A Review of Traditional Technologies Based on Literature and Field Experiences*. FAO, Rome, Italy.

Forestry Commission, Scotland (2016). Woodland Grazing Toolbox. http://scotland.forestry.gov.uk/woodland-grazing-toolbox/grazing-management/foraging/palatability-and-resilience-of-native-trees (accessed on 23 April 2017).

Franzluebbers, A. J. (2007). Integrated crop-livestock systems in the southeastern USA. *Agronomy Journal*, 99, pp. 361-72.

Gallardo, A. (2003). Effect of tree canopy on the spatial distribution of soil nutrients in a Mediterranean dehesa. *Pedobiologia*, 47(2), pp. 117-25.

Garrity, D. P. (2004). Agroforestry and the achievement of the Millennium Development Goals. *Agroforestry Systems*, 61(1), pp. 5-17.

Garrity, D., Okono, A., Grayson, M. and Parrott, S. (2006). *World Agroforestry into the Future*. World Agroforesty Centre, Nairobi.

Gea-Izquierdo, G. and Cañellas, I. (2009). Analysis of holm oak intraspecific competition using gamma regression. *Forest Science*, 55(4), pp.310-22.

Gea-Izquierdo, G., Montero, G. and Cañellas, I. (2009). Changes in limiting resources determine spatio-temporal variability in tree-grass interactions. *Agroforestry Systems*, 76(2), pp. 375–87.

Gibon, A., Sheeren, D., Monteil, C., Ladet, S. and Balent, G. (2010). Modelling and simulating change in reforesting mountain landscapes using a social-ecological framework. *Landscape Ecology*, 25(2), pp. 267–85.

Gibon, A., Ryschawy, J., Schaller, N., Fiorelli, J. L., Havet, A. and Martel, G. (2012). Why and how to analyse the potential of mixed crop-livestock farming systems for sustainable agriculture and rural development at landscape level. In: *Producing and Reproducing Farming Systems. New Modes of Organization for Sustainable Food Systems of Tomorrow. 10th European IFSA Symposium, Aarhus, Denmark, 1–4 July 2012*. International Farming Systems Association, Vienna, pp. unpaginated ref.30.

Gillespie, A. R., Miller, B. K. and Johnson K. D. (1995). Effects of ground cover on tree survival and growth in filter strips of the Corn Belt region of the Midwestern U.S. *Agriculture, Ecosystems and Environment*, 53, pp. 263–70.

Giraldo, C., Escobar, F., Chará, J. D. and Calle, Z. (2011). The adoption of silvopastoral systems promotes the recovery of ecological processes regulated by dung beetles in the Colombian Andes. *Insect Conservation and Diversity*, 4, pp. 115–22.

Gómez-Gutiérrez, J. M. and Pérez-Fernández, M. (1996). The 'dehesas': Silvopastoral systems in semiarid Mediterranean regions with poor soils, seasonal climate and extensive utilization. In: M. Etienne (Ed.), *Western European Silvopastoral Systems*. INRA Editions, Paris, pp. 55–70.

Gregory, N. G. (1997). The role of shelterbelts in protecting livestock: A review. *New Zealand Journal of Agricultural Research*, 38, pp. 423–50.

Gupta, S. C., Gupta, N. and Nivsarkar, A. E. (1999). *Mithun – A Bovine of Indian Origin*. Indian Council of Agricultural Research, New Delhi. Referenced in: Phillips, C. J. C. (2002). *Cattle Behaviour and Welfare*. Blackwell Science Ltd. Oxford, UK.

Haile, S. G., Nair, V. D. and Nair, P. K. R. (2010). Contribution of trees to carbon storage in soils of silvopastoral systems in Florida, USA. *Global Change Biology*, 16(1), pp. 427–38.

Halvorson, J. J., Cassida, K. A., Turner, K. E. and Belesky, D. P. (2011). Nutritive value of bamboo as browse for livestock. *Renewable Agriculture and Food Systems*, 26, 161–70.

Hart, S. (2013). Identification and management of different browse species adapted to the Southern region. In: *Sustainable Year-Round Forage Production and Grazing/Browsing Management for Goats in the Southern Region*. Southern SARE, Tuskegee University Extension, Tuskegee, Alabama, pp. 123–30.

Hemery, G. E., Savill, P. S. and Pryor, S. N. (2005). Applications of the crown diameter-stem diameter relationship for different species of broadleaved trees. *Forest Ecology and Management*, 215, pp. 285–94.

Hendrickson, J. and Olson, B. (2006). Understanding plant response to grazing. In: K. Launchbaugh (Ed.), *Targeted Grazing: A Natural Approach to Vegetation Management and Landscape Enhancement*. American Sheep Industry Association and Cottrell Printing, Centennial, CO, pp. 32–9.

Hendrickson, J. R., Hanson, J. D., Tanaka, D. L. and Sassenrath, G. (2008). Principles of integrated agricultural systems: Introduction to processes and definition. *Renewable Agriculture and Food Systems*, 23(4), pp. 265–71.

Henkin, Z., Ungar, E. D., Dvash, L., Perevolotsky, A., Yehuda, Y., Sternberg, M., Voet, H. and Landau, S. Y. (2011). Effects of cattle grazing on herbage quality in a herbaceous Mediterranean rangeland. *Grass and Forage Science*, 66, pp. 516–25.

Hillerton J. E., Bramley A. J. and Broom D. M. (1983). *Hydrotaea irritans* and summer mastitis in calves. *Veterinary Record*, 113, pp. 88-9.

Howlett, D. S., Moreno, G., Mosquera Losada, M. R., Nair, P. K. R. and Nair, V. D. (2011). Soil carbon storage as influenced by tree cover in the dehesa cork oak silvopasture of central-western Spain. *Journal of Environmental Monitoring*, 13(7), pp. 1897–904.

Huzzey, J. M., Weary, D. M. and Von Keyserlingk, M. A. G. (2013). Changes in exploratory feeding behaviour as an early indicator of metritis in dairy cattle. In: M. J. Hötzel and L. C. P. M. Filho (Eds), *Proceedings of 47th Congress of International Society for Applied Ethology*, 2-6 June 2013, Florianópolis, Brazil, p. 107.

IFAD (2010). *Integrated Crop-livestock Farming Systems*, International Fund for Agricultural Development, Rome, Italy.

IFOAM (2005). The principles of organic agriculture. Available at: http://www.ifoam.bio/en/organic-landmarks/principles-organic-agriculture (accessed 22 October 2016).

Intergovernmental Panel on Climate Change (IPCC) (2007). *Climate Change 2007: The Physical Life Science Basis. Group I Contribution to the Fourth Assessment Report of the IPCC*. Cambridge University Press, Cambridge, UK and New York, NY.

Janiszewski, P., Bogdaszewski, M., Murawska, D. and Tajchman, K. (2016). Welfare of farmed deer - practical aspects. *Polish Journal of Natural Sciences*, 31, pp. 345-61.

Joffre, R. and Rambal, S. (1988). Soil-water improvement by trees in the rangelands of southern Spain. *Acta Oecologica*, 9(4), pp. 405–22.

Jones, T., Feber, R., Hemery, G., Cook, P., James, K., Lamberth, C. and Dawkins, M. (2007). Welfare and environmental benefits of integrating commercially viable free-range broiler chickens into newly planted woodland: A UK case study. *Agricultural Systems*, 94, pp. 177–88.

Jose, S. (2005). Ecological interactions in silvopastoral systems: A synthesis. *Annals of Arid Zone Journal*, 44, pp. 327–36.

Jose, S. (2009). Agroforestry for ecosystem services and environmental benefits: An overview. *Agroforestry Systems*, 34, pp. 27-31.

Jose, S., Gillespie, A. R. and Pallardy, S. G. (2004). Interspecific interactions in temperate agroforestry. *Agroforestry Systems*, 61, pp. 237–55.

Karki, U. and Goodman, M. S. (2009). Cattle distribution and behavior in southern-pine silvopasture versus open-pasture. *Agroforestry Systems*, 78, pp. 159–68.

Kearney, P. E., Murray, P. J., Hoy, J. M., Hohenhaus, M. and Kotze, A. (2016). The 'Toolbox' of strategies for managing Haemonchus controtus in goats: What's in and what's out. *Veterinary Parasitology*, 220, pp. 93–107.

Kemp, P. D., Barry, T. N. and Douglas, G. B. (2003). Edible forage yield and nutritive value of poplar and willow. In: *Using Trees on Farms*. Grassland Research and Practice Series No. 10, pp. 53–63.

Kerslake, J. I., Everett-Hincks, J. M. and Campbell A. W. (2005). Lamb survival: A new examination of an old problem. *Proceedings of the New Zealand Society of Animal Production*, 65, pp. 13–18.

Kremen, C. and Miles, A. (2012). Ecosystem services in biologically diversified versus conventional farming systems: Benefits, externalities, and trade-offs. *Ecology and Society*, 17(4), p. 40.

Kremen, C., Iles, A. and Bacon, C. (2012). Diversified farming systems: An agroecological, system-based alternative to modern industrial agriculture. *Ecology and Society*, 17(4), p. 44.

Laister, S., Stockinger, B., Regner, A. M., Zenger, K., Knierim, U. and Winckler, C. (2011). Social licking in dairy cattle: Effects on heart rate in performers and receivers. *Applied Animal Behaviour Science*, 130, 81-90.

Larsen, H., Cronin, G., Smith, C. L., Hemsworth, P. and Rault, J.-L. (2017). Behaviour of free-range hens in distinct outdoor environments. *Animal Welfare*, 26, pp. 255-64.

Le Houérou, H. N. (2000). Utilization of fodder trees and shrubs in the arid and semiarid zones of West Asia and North Africa. *Arid Soil Research and Rehabilitation*, 14, pp. 101-35.

Lemaire, G., Franzluebbers, A., Cesar de Faccio Carvalho, P. and Dedieu, B. (2014). Integrated crop-livestock systems: Strategies to achieve synergy between agricultural production and environmental quality. *Agriculture, Ecosystems and Environment*, 190, pp. 4-8.

Lidfors, L. M., Moran, D., Jung, J., Jensen, P. and Castren, H. (1994). Behaviour at calving and choice of calving place in cattle kept in different environments. *Applied Animal Behaviour Science*, 42, pp. 11-28.

Liu, B., Tu, C., Hu, S., Gumptertz, M. and Ristaino, J. B. (2007). Effect of organic, sustainable, and conventional management strategies in grower fields on soil physical, chemical, and biological factors and the incidence of Southern blight. *Applied Soil Ecology*, 37(3), pp. 202-14.

Liu, T., Rodríguez, L. F., Green, A. R., Shike, D. W., Segers, J. R., Del Nero Maia, G. and Norris, H. D. (2012). Assessment of cattle impacts on soil characteristics in integrated crop-livestock systems. In: *Abstracts from the American Society of Agricultural and Biological Engineers Annual International Meeting*, Dallas.

Luske, B. and van Eekeren, N. (2014). Renewed interest for silvopastoral systems in Europe: An inventory of the feeding value of fodder trees. In *4th ISOFAR Scientific Conference*. Istanbul, Turkey, 13-15 October 2014, pp. 811-14.

Luske, B. and van Eekeren, N. (2015). Potential of fodder trees in high-output dairy systems. In: A. van den Pol-van Dasselaar, H. F. M. Aarts, A. De Vliegher, A. Elgersma, D. Reheul, J. A. Reijneveld, J. Verloop and A. Hopkins (Eds), *Grassland Science in Europe, Vol. 20. Grassland and Forages in High Output Dairy Farming Systems*, pp. 250-2.

Luttikholt, L. (2007). Principles of organic agriculture as formulated by the International Federation of Organic Agriculture Movements. *NJAS - Wageningen Journal of Life Sciences*, 54(4), pp. 347-60.

Maia, M. F. and Moore, S. J. (2011). Plant-based insect repellents: A review of their efficacy, development and testing. *Malaria Journal*, 10(Suppl. 1), pp. S11. http://www.malariajournal.com/content/10/S1/S11

Mancera, A. K. and Galindo, F. (2011). Evaluation of some sustainability indicators in extensive bovine stockbreeding systems in the state of Veracruz. VI Reunión Nacional de Innovación Forestal 31. León Gua najauato, México. Referenced in: Broom, D. M. (2017). Components of sustainable animal production and the use of silvopastoral systems. *Revista Brasileira de Zootecnia*, 46, pp. 683-8.

Marchão, R. L., Balbino, L. C., Da Silva, E. M., Dos Santos Jr., J. D. D. G., De Sá, M. A. C., Vilela, L. and Becquer, T. (2007). Soil physical quality under crop-livestock management systems in a Cerrado Oxisol. *Pesquisa Agropecuária Brasileira*, 42(6), pp. 873-82.

Marchão, R. L., Lavelle, P., Celini, L., Balbino, L. C., Vilela, L. and Becquer, T. (2009). Soil macrofauna under integrated crop-livestock systems in a Brazilian Cerrado Ferralsol. *Pesquisa Agropecuária Brasileira*, 44(8), pp. 1011-20.

Martha Jr., G. B., Alves, E. and Contini, E. (2011). Economic dimension of integrated crop-livestock systems. *Pesquisa Agropecuária Brasileira*, 46, pp. 1117-26.

McAdam, J. and McEvoy, P. M. (2009). The potential for silvopastoralism to enhance biodiversity on grassland farms in Ireland. In: A. Rigueiro-Rodríguez, J. McAdam and M. Mosquera-Losada (Eds), *Agroforestry in Europe: Current Status and Future Prospects*. Dordrecht, the Netherlands: Springer Science + Business Media B.V., pp. 19-24.

McAdam, J. H., Sibbald, A. R., Teklehaimanot, Z. and Eason, W. R. (2007). Developing silvopastoral systems and their effects on diversity of fauna. *Agroforestry Systems*, 70(1), pp. 81-9.

McCutcheon, S. N., Holmes, C. W. and McDonald, M. F. (1981). The starvation-exposure syndrome and neonatal lamb mortality: A review. *Proceedings of the New Zealand Society of Animal Production*, 47, pp. 209-17.

McDonald, P., Edwards, R. A., Greenhalgh, J. F. D. and Morgan, C. A. (1995). *Animal Nutrition*, 5th ed. Addison Wesley Longman Limited, Harlow, UK.

Menegaz De Farias, P., Arellano, L., Hernández, M. I. M. and Ortiz, S. L. (2015). Response of the copro-necrophagous beetle (Coleoptera: Scarabaeinae) assemblage to a range of soil characteristics and livestock management in a tropical landscape. *Journal of Insect Conservation*. DOI:10.1007/s10841-015-9812-3.

Min, B. R. and Hart, S. P. (2003). Tannins for suppression of internal parasites. *Journal of Animal Science*, 81 (E Suppl. 2), pp. E102-9.

Min, B. R., Barry, T. N., Attwood, G. T. and McNabb, W. C. (2003). The effect of condensed tannins on the nutrition and health of ruminants fed fresh temperate forages: A review. *Animal Feed Science and Technology*, 106, pp. 3-19.

Minnaar, I. A., Bennett, N. C., Chiminba, T. and McKechnie, A. E. (2014). Summit metabolism and metabolic expansibility in Wahlberg's epauletted fruit bats (*Epomorphorus wahlbergi*): Seasonal acclimatisation and effects of captivity. *The Journal of Experimental Biology*, 217, pp. 1363-9.

Mitlohner, F. M., Morrow, J. L., Dailey, J. W., Wilson, S. C., Galyean, M. L., Miller, M. F. and McGlone, J. J. (2001). Shade and water misting effects on behavior, physiology, performance, and carcass traits of heat-stressed feedlot cattle. *Journal of Animal Science*, 79, pp. 2327-35.

Montagnini, F., Ibrahim, M. and Restrepo, E. M. (2013). Silvopastoral systems and climate change mitigation in Latin America. *Bois et Forèts des Tropiques*, 316, pp. 3-16.

Moore, K. M., Barry, T. N., Cameron, P. N., Lopez-Villalobos, N. and Cameron, D. J. (2003). Willow (*Salix* sp.) as a supplement for grazing cattle under drought conditions. *Animal Feed Science and Technology*, 104, pp. 1-11.

Mooring, S. M. and Samuel, W. M. (1998). Tick defense strategies in bison: The role of grooming and hair coat. *Behaviour*, 135, pp. 693-718.

Moraine, M., Duru, M., Nicholas, P., Leterme, P. and Therond, O. (2014). Farming system design for innovative crop-livestock integration in Europe. *Animal*, 8, pp. 1204-7.

Moraine, M., Duru, M. and Therond, O. (2016). A social-ecological framework for analyzing and designing integrated crop-livestock systems from farm to territory levels. *Renewable Agriculture and Food Systems*, 1, pp.1-15.

Moreno, G. and Obrador, J. J. (2007). Impact of evergreen oaks on soil fertility and crop production in intercropped dehesas. *Agriculture, Ecosystems & Environment*, 119(3-4), pp. 270-80.

Moreno, G., Obrador, J. J. and Garcia, A. (2007). Impact of evergreen oaks on soil fertility and crop production in intercropped dehesas. *Agriculture, Ecosystems & Environment*, 119(3-4), pp. 270-80.

Morgan-Davies, J., Morgan-Davies, C., Pollock, M. L., Holland, J. P. and Waterhouse, A. (2014). Characterisation of extensive beef cattle systems: Disparities between opinions, practice and policy. *Land Use Policy*, 38, pp. 707-18.

Mosquera-Losada, M. R., McAdam, J. H., Romero-Franco, R., Santiago-Freijanes, J. J. and Rigueiro-Rodríguez, A. (2009). Definitions and components of agroforestry practices in Europe. In: A. Rigueiro-Rodríguez, J. McAdam and M. R. Mosquera-Losada (Eds), *Agroforestry in Europe: Current status and future prospects*. Dordrecht, the Netherlands: Springer Science + Business Media B.V., pp. 3-19.

Mutuo, P., Cardisch, G., Albrecht, A., Palm, C. A. and Verchot, L. (2005). Potential of agroforestry for carbon sequestration and mitigation of greenhouse gas emissions from soils in the tropics. *Nutrient Cycling in Agroecosystems*, 71(1), pp. 43-54.

Muwanga, S., Mugisha, A. and Vaarst, M. (2011). Organic livestock production in Uganda: Potentials, challenges and prospects. *Tropical Animal Health and Production*, 43(4), pp. 749-57.

Nair, P. K. (2011). Agroforestry systems and environmental quality: Introduction. *Journal of Environmental Quality*, 40, pp. 784-90.

Nair, V. D. and Graetz, D. A. (2004). Agroforestry as an approach to minimizing nutrient loss from heavily fertilized soils: The Florida experience. *Agroforestry Systems*, 61, pp. 269-79.

Nair, V. D., Nair, P. K. R., Kalmbacher, R. S. and Ezenwa, I. V. (2007). Reducing nutrient loss from farms through silvopastoral practices in coarse-textured soils of Florida, USA. *Ecological Engineering*, 29(2), pp. 192-9.

Nair, P. K. R., Kumar, B. M. and Nair, V. D. (2009). Agroforestry as a strategy for carbon sequestration. *Journal of Plant Nutrition and Soil Science*, 172(1), pp. 10-23.

Nardone, A., Ronchi, B., Lacetera, N., Ranieri, M. S. and Bernabucci, U. (2010). Effects of climate changes on animal production and sustainability of livestock systems. *Livestock Science*, 130(1-3), pp.57-69.

Novobilský, A., Mueller-Harvey, I. and Thamsborg, S. M. (2011). Condensed tannins act against cattle nematodes. *Veterinary Parasitology*, 182, pp. 213-20.

Núñez, V., Hernando, A. and Velázquez Tejera, R. (2012). Livestock management in Natura 2000: A case study in a *Quercus pyrenaica* neglected coppice forest. *Journal for Nature Conservation*, 20(1), pp. 1-9.

Ocampo, A., Cardozo, A., Tarazona, A., Ceballos, M. and Murgueitio, E. (2011). La investigación participativa en bienestar y comportamiento animal en el trópico de América: Oportunidades para nuevo conocimiento aplicada. *Revista Colombiana Ciencias Pecuarias*, 24, pp. 332-46.

Oliveira, S. E. O., Maia, A. S. C., Costa, C. M. and Neto, M. C. (2013). Behavioural indicators of cattle welfare in silvopastoral systems in the tropics of México. In: M. J. Hötzel and L. C. P. M. Filho (Eds), *Proceedings of 47th Congress of International Society for Applied Ethology*, 2-6 June 2013, Florianópolis, Brazil, p. 128.

Orefice, J. N. (2015). Silvopasture in the Northeastern United States. PhD Thesis. University of New Hampshire, Durham, NH. www.unh.edu

Palma, J. H. N., Graves, A. R., Bunce, R. G. H., Burgess, P. J., de Filippo, R., Keesman, K. J., van Keulen, H., Liagre, F., Mayus, M., Moreno, G., Reisner, Y. and Herzog, F. (2007). Modelling environmental benefits of silvoarable agroforesty in Europe. *Agriculture, Ecosystems & Environment*, 119(3-4), pp. 320-34.

Papachristou, T. G., Dziba, L. E., Provenza, F. D. (2005). Foraging ecology of goats and sheep on wooded rangelands. *Small Ruminant Research*, 59, pp. 141-56.

Pent, G. J. (2017). Lamb performance, behavior, and body temperatures in hardwood silvopasture systems. PhD Thesis. Virginia Polytechnic Institute and State University.

Percival, N. S., Hawke M. F. and Andrew B. L. (1984a). Preliminary report on climate measurements under radiata pine planted on farmland. In: G. W. Bilbrough (Ed.), *Proceedings of a Technical Workshop on Agroforestry*. Ministry of Agriculture and Fisheries, Dunedin, New Zealand, pp. 57-60.

Percival, N. S., Hawke, M. F., Bond, D. I. and Andrew B. L. (1984b). Livestock carrying capacity and performance on pasture under *Pinus radiata*. *Proceedings of a Technical Workshop on Agroforestry*. Ministry of Agriculture and Fisheries, Dunedin, New Zealand, pp. 25-31.

Perrot, C., Reuillon, J., Capitain, M., et al. (2008). Phasing out milk quotas: Strengths and weaknesses of French dairy farms in mountain areas. *Rencontre Recherche Ruminants*, 15, pp. 243-6.

Phillips, C. J. C. (1993). *Cattle Behaviour*. Farming Press Books, Ipswich.

Pitta, D. W., Barry, T. N., Lopez-Villalobos, N. and Kemp, P. D. (2004). The effect of sheep reproduction of grazing willow fodder blocks during drought. *Proceedings of the New Zealand Society of Animal Production*, 64, pp. 67-71.

Plieninger, T. (2007). Compatibility of livestock grazing with stand regeneration in Mediterranean holm oak parklands. *Journal for Nature Conservation*, 15, pp.1-9.

Poudel, S., Karki, U., McElhenney, W., Karki, Y., Tillman, A., Karki, L. and Kumi, A. (2017). Challenges of stocking small ruminant in grazing plots with dormant browse species. *Professional Agricultural Workers Journal*, 5, pp. 28-35.

Price, P. W., Bouton, C. E., Gross, P., McPherson, B. A., Thompson, J. N. and Weis, A. E. (1980). Interactions among three trophic levels: Influence of plants on interactions between insect herbivores and natural enemies. *Annual Review of Ecology and Systematics*, 11, pp. 41-65.

Provenza, F. D. (1995). Postingestive feedback as a elementary determinant of food preference and intake in ruminants. *Journal of Rangeland Management*, 48, pp. 2-17.

Provenza, F. D., Meuret, M. and Gregorini, P. (2015). Our landscapes, our livestock, ourselves: Restoring broken linkages among plants, herbivores, and humans with diets that nourish and satiate. *Appetite*. 95, pp. 500-19.

Pulido-Santacruz, P. and Renjifo, L. M. (2011). Live fences as tools for biodiversity conservation: A study case with birds and plants. *Agroforestry Systems*, 81, pp. 15-30.

Radostits, O. M., Gay, C. C., Hinchcliff, K. W. and Constable, P. D. (2007). *Veterinary Medicine. A Textbook of the Diseases of Cattle, Sheep, Goats, Pigs and Horses*. WB Saunders, Philadelphia, PA.

Ramírez-Restrepo, C. A., Barry, T. N., López-Villalobos, N, Kemp, P. D. and Harvey, T. G. (2005). Use of *Lotus corniculatus* containing condensed tannins to increase efficiency in ewes under commercial dryland farming conditions. *Animal Feed Science and Technology*, 121, pp. 23–43.

Riedel, J. L., Bernués A and Casasús I (2013). Livestock grazing impacts on herbage and shrub dynamics in a Mediterranean Natural Park. *Rangeland Ecology & Management*, 66(2), pp. 224–33.

Rigueiro-Rodríguez, A., Mosquera-Losada, M. R. and López-Díaz, M. L. (2008). Effect of sewage sludge and liming on productivity during the establishment of a silvopastoral system in north-west Spain. *New Zealand Journal of Agricultural Research*, 51(2), pp. 199–207.

Robertson, H., Marshall, D., Slingsby, E. and Newman, G. (2012a). Economic, biodiversity, resource protection and social values of orchards: A study of six orchards by the Herefordshire Orchards Community Evaluation Project. Natural England Commissioned Reports, Number 090. http://publications.naturalengland.org.uk/publication/1289011

Robertson, S. M., King, B. J., Broster, J. C. and Friend, M. A. (2012b). The survival of lambs in shelter declines at high stocking intensities. *Animal Production Science*, 52, pp. 497–501.

Robinson, B. H., Mills, T. M., Green, S. R., Chancerel, B., Clothier, B. E., Fung, L., Hurst, S. and McIvor, I. (2005). Trace element accumulation by poplars and willows used for stock fodder. *New Zealand Journal of Agricultural Research*, 48, pp. 489–97.

Ronchi, B. and Nardone, A. (2003). Contribution of organic farming to increase sustainability of Mediterranean small ruminants livestock systems. *Livestock Production Science*, 80(1–2), pp. 17–31.

Rozados-Lorenzo, M. J., González-Hernández, M. P. and Silva-Pando, F. J. (2007). Pasture production under different tree species and densities in an Atlantic silvopastoral system. *Agroforestry Systems*, 70, pp. 53–62.

Russelle, M. P., Entz, M. H. and Franzluebbers, A. J. (2007). Reconsidering integrated crop-livestock systems in North America. *Agronomy Journal*, 99, pp. 325–34

Ryschawy, J., Choisis, N., Choisis, J. P. and Gibon, A. (2012). Mixed crop-livestock systems: An economic and environmental-friendly way of farming? *Animal*, 6(10), pp. 1722–30.

Ryschawy, J., Choisis, N., Choisis, J. P. and Gibon, A. (2013). Paths to last in mixed crop-livestock farming: Lessons from an assessment of farm trajectories of change. *Animal*, 7(4), pp. 673–8.

Ryschawy, J., Joannon, A., Choisis, J. P. and Le Gal, P.-Y. (2014a). Participative assessment of innovative technical scenarios for enhancing sustainability of French mixed crop-livestock farms. *Agricultural Systems*, 129, pp. 1–8.

Ryschawy, J., Alexandre, J. and Gibon, A. (2014b). Mixed crop-livestock farm: Definitions and research issues. A review. *Cahiers Agricultures*, 23(6), pp. 346–56.

Ryschawy, J., Martin, G., Moraine, M. and Therond, O. (2015). Designing crop-livestock integration at different levels: Toward new agroecological models? *Nutrient Cycling in Agroecosystems*, 108(1), 5–20.

Sanderson, M. A., Archer, D., Hendrickson, J., Kronberg, S., Liebig, M., Nichols, K., Schmer, M., Tanaka, D. and Aguilar, J. (2013). Diversification and ecosystem services for conservation agriculture: Outcomes from pastures and integrated crop-livestock systems. *Renewable Agriculture and Food Systems*, 28(2), pp. 129-44.

Sanga, U., Provenza, F. D. and Villalba, J. J. (2011). Transmission of self-medicative behaviour from mother to offspring in sheep. *Animal Behaviour*, 82, pp. 219-27.

Santos, G. G., Marchão, R. L., da Silva, E. M., da Silveira, P. M. and Becquer, T. (2011). Soil physical quality in integrated crop-livestock systems. *Pesquisa Agropecuária Brasileira*, 46(19), pp. 1339-48.

Šárová, R., Gutmann, A. K., Špinka, M., Stěhulová, I. and Winckler, C. (2016). Important role of dominance in allogrooming behaviour in beef cattle. *Applied Animal Behaviour Science*, 181, pp. 41-8.

Schädler, M., Brandl, R. and Kempel, A. (2010). 'Afterlife' effects of mycorrhization on the decomposition of plant residues. *Soil Biology and Biochemistry*, 42(3), pp. 521-3.

Schmidt, M. H. and Tscharntke, T. (2005). The role of perennial habitats for Central European farmland spiders. The role of perennial habitats for Central European farmland spiders. *Agriculture, Ecosystems & Environment*, 105(1-2), pp.235-42.

Schütz, K. E., Rogers, A. R., Poulouin, Y. A., Cox, N. R. and Tucker, C. B. (2010). The amount of shade influences the behavior and physiology of dairy cattle. *Journal of Dairy Science*, 93, pp. 125-33.

Seffan-Dewenter, I. and Leschke, K. (2003). Effects of habitat management on vegetation and above-ground nesting bees and wasps of orchard meadows in Central Europe. *Biodiversity and Conservation*, 12, pp. 1953-68.

Segnalini, M., Bernabucci, U., Vitali, A., Nardone, A. and Lacetera, N. (2013). Temperature humidity index scenarios in the Mediterranean basin. *International Journal of Biometeorology*, 57(3), pp. 451-8.

Shashua-Bar, L., Pearlmutter, D. and Erell, E. (2009). The cooling efficiency of urban landscape strategies in a hot dry climate. *Landscape and Urban Planning*, 92, pp. 179-86.

Shipley, L. A. (1999). Grazers and browsers: How digestive morphology affects diet selection. In: K. L. Launchbaugh, K. D. Sanders and J. C. Mosley (Eds), *Grazing Behavior of Livestock and Wildlife*. University of Idaho, 20-7. Idaho Forest, Wildlife and Range Experimental Station Bulletin no. 70.

Shurkin, J. (2014). Animals that self-medicate. *Proceedings of the National Academy of Science*, 111, pp. 17339-41.

Silanikove, N. (2000). Effects of heat stress on the welfare of extensively managed domestic ruminants. *Livestock Production Science*, 67, pp. 1-18.

Simón, N., Montes, F., Díaz-Pinés, E., Benavides, R., Roig, S. and Rubio, A. (2013). Spatial distribution of the soil organic carbon pool in a Holm oak dehesa in Spain. *Plant and Soil*, 366(1), pp. 537-49.

Sinclair, F. L., Eason, W. R. and Hooker, J. (2000). Understanding and management of interactions. In: A. M., Hislop and J. Claire (Eds), *Agroforestry in the UK*. Forestry Commission, Edinburgh, UK, pp. 17-28.

Smith, J., Pearce, B. D. and Wolfe, M. S. (2012). A European perspective for developing modern multifunctional agroforestry systems for sustainable intensification. *Renewable Agriculture and Food Systems*, 27(4), pp. 323-32.

Smith, J., Pearce, B. D. and Wolfe, M. S. (2013). Reconciling productivity with protection of the environment: Is temperate agroforestry the answer? *Renewable Agriculture and Food Systems*, 28(1), pp. 80-92.

Smith, J., Whistance, L., Costanza, A. and Demeretz, V. (2017). System report: Agroforestry for ruminants in England. Agroforestry for livestock farmers. Contribution to Deliverable 5.14: Lessons learnt from innovations related to agroforestry for livestock for EU. F.P7 Research Project: AGFORWARD 613520. 20 pp. http://www.agforward.eu/index.php/en/agroforestry-with-ruminants-uk.html

Solorio Sanchez, F. J. and Solorio Sanchez, B. (2002). Integrating fodder trees in to animal production in the tropics. *Tropical and Subtropical Agroecosystems*, 1, pp. 1-11.

Sommerville, R. and Jones, T. (2013). Achieving a high quality of life for farm animals requires provision rather than deprivation. In: M. J. Hötzel and L. C. P. M. Filho (Eds), *Proceedings of 47th Congress of International Society for Applied Ethology*, 2-6 June 2013, Florianópolis, Brazil, p. 54.

Soussana, J.-F. and Lemaire, G. (2014). Coupling carbon and nitrogen cycles for environmentally-sustainable intensification of grasslands and crop-livestock systems. *Agriculture, Ecosystems and Environment*, 190, pp. 9–13.

SSBA (2008). Two crops from one acre: A comprehensive guide to using Shropshire sheep for grazing tree plantations. *Shropshire Sheep Breeders' Association*. www.shropshire-sheep.co.uk.

Sulc, R. and Tracy, B. (2007). Integrated crop-livestock systems in the US corn belt. *Agronomy Journal*, 99, pp. 335–45.

Sundrum, A. (2001). Organic livestock farming: A critical review. *Livestock Production Science*, 67, pp. 207–15.

Taiz, L. and Zeiger, E. (2010). *Plant Physiology*, 5th ed. Sinauer Associates Inc., Sunderland, MA.

Tamang, B., Andreu, M. G., Friedman, M. H. and Rockwood, D. L. (2009). Windbreak designs and planting for Florida agricultural fields. FOR227. University of Florida Institute of Food and Agricultural Sciences, Gainesville, FL. http://edis.ifas.ufl.edu/fr289 (Accessed on 12 September 2017).

Tarawali, S., Herrero, M., Descheemaeker, K., Grings, E. and Blümmel, M. (2011). Pathways for sustainable development of mixed crop livestock systems: Taking a livestock and pro-poor approach. *Livestock Science*, 139(1–2), pp. 11–21.

Tárrega, R., Calvo, L., Taboada, A., García-Tejero, S. and Marcos, E. (2009). Abandonment and management in Spanish dehesa systems: Effects on soil features and plant species richness and composition. *Forest Ecology and Management*, 257(2), pp. 731–8.

The International Assessment of Agricultural Knowledge, Science and Technology for Development (2008). Executive summary of the synthesis report. Available at: http://www.agassessment.org/docs/IAASTD_EXEC_SUMMARY_JAN_2008.pdf (accessed on 03 December 2014).

Thevathasan, N. V. and Gordon, A. M. (2004). Ecology of tree intercropping systems in the North temperate region: Experiences from southern Ontario, Canada. *Agroforestry Systems*, 61(1), pp. 257–68.

Thorhallsdottir, A. G., Provenza, F. D. and Balph, D. F. (1990). Ability of lambs to learn about novel foods while observing or participating with social models. *Applied Animal Behaviour Science*, 25, pp. 25–33.

Udawatta, R. P., Kremer, R. J., Adamson, B. W. and Anderson, S. H. (2008). Variations in soil aggregate stability and enzyme activities in a temperate agroforestry practice. *Applied Soil Ecology*, 39(2), pp.153–60.

Vaarst, M. (2015). The role of animals in eco-functional intensification of organic agriculture. *Sustainable Agriculture Research*, 5(3), pp. 103–15.

Vandermeulen, S., Ramírez-Restrepo, C. A., Marche, C., Decruyenaere, V., Beckers, Y. and Bindelle, J. (2016). Behaviour and browse species selectivity of heifers grazing in a temperate silvopastoral system. *Agroforestry Systems*. doi:10.1007/s10457-016-0041-x.

Van Keulen, H. and Schiere, H. (2004). Crop-livestock systems: Old wine in new bottles?. In: Anon. (Ed.), *New Directions for a Diverse Planet. Proceedings of the 4th International Crop Science Congress*, Brisbane, Australia.

Veysset, P., Lherm, M., Roulenc, M., Troquier, C. and Bebin, D. (2015). Productivity and efficiency of French suckler beef production systems: Trends over the last 20 years. In: *Annual Meeting of the European Federation of Animal Science*. Book of Abstracts No. 21, Warsaw.

Vilela, L., Martha Jr., G. B., Macedo, M. C. M., Marchão, R. L., Guimarães Jr., R., Pulrolnik, K. and Maciel, G. A. (2011). Integrated crop-livestock systems in the cerrado region. *Pesquisa Agropecuária Brasileira*, 46(10), pp. 1127–38.

Villalba, J. J. and Provenza, F. D. (2007). Self-medication and homeostatic behaviour in herbivores: Learning about the benefits of nature's pharmacy. *Animal*, 1, pp. 1360–70.

Villalba, J. J. and Provenza F. D. (2009). Learning and dietary choice in herbivores. *Rangeland Ecology and Management*, 62, pp. 399–406.

Villalba, J. J., Provenza, F. D. and Shaw, R. (2006). Sheep self-medicate when challenged with illness-inducing foods. *Animal Behaviour*, 71, pp. 1131–9.

Villalba, J. J., Provenza, F. D. and Hall, J. O. (2008). Learned appetites for calcium, phosphorus, and sodium in sheep. *Journal of Animal Science*, 86, pp. 738–47.

Volk, T. A., Abrahamson, L. P., Nowak, C. A., Smart, L. B., Tharakan, P. J. and White, E. H. (2006). The development of short-rotation willow in the northeastern United States for bioenergy and bioproducts, agroforestry and phytoremediation. *Biomass and Bioenergy*, 30(8–9), pp. 715–27.

Waghorn, G. C. and McNabb, W. C. (2003). Consequences of plant phenolic compounds for productivity and health of ruminants. *Proceedings of the Nutrition Society*, 62, pp. 383–92.

Walkden-Brown, S. W. and Kahn, L. P. (2002). Nutritional modulation of resistance and resilience to gastrointestinal nematode infection - a review. *Asian-Australian Journal of Animal Science*, 15, pp. 912–24.

Waller, P. J., Bernes, G., Thamsborg, S. M., Sukura, A., Richter, S. H., Ingebrigtsen, K. and Höglund, J. (2001). Plants as de-worming agents of livestock in the Nordic countries: Historical perspective, popular beliefs and prospects for the future. *Acta Veterinaria Scandinavica*, 42, pp. 31–44.

Wass, J. A., Pollard, J. C. and Littlejohn, R. P. (2004). Observations on the hiding behaviour of farmed red deer (*Cervus elaphus*) calves. *Applied Animal Behaviour Science*, 88, pp. 111–20.

Wick, B. and Tiessen, H. (2008). Organic matter turnover in light fraction and whole soil under silvopastoral land use in semiarid northeast Brazil. *Rangeland Ecology & Management*, 61(3), pp. 275–83.

Wiersma, P., Chappell, M. A. and Williams, J. B. (2007). Cold- and exercise-induced peak metabolic rates in tropical birds. *Proceedings of the National Academy of Science*, 104, pp. 20866–71.

Wilkins, R. J. (2008). Eco-efficient approaches to land management: A case for increased integration of crop and animal production systems. *Philosophical Transactions of the Royal Society of London. Series B, Biological Sciences*, 363(1492), pp. 517–25.

Williams, A. R., Fryganas, C., Ramsay, A., Mueller-Harvey, I. and Thamsborg, S. M. (2014). Direct anthelmintic effects of condensed tannins from diverse plant sources against *Ascaris suum*. *PLoS ONE*, 9(5), p. e97053. doi:10.1371/journal.pone.0097053.

Wilson, A. D. (1969). A review of browse in the nutrition of grazing animals. *Journal of Rangeland Management*, 22, pp. 23–8.

Younie, D. (2000). Integration of livestock into organic farming systems: Health and welfare problems. In: Hovi, M. and Garcia Trujillo, R. (Eds), *Proceedings of the Second NAHWOA Workshop*, Cordoba, Spain.

# Chapter 2

## Tree planting and management in agroforestry

*Lydie Dufour, INRA, France*



### 1 Introduction

Agroforestry is currently being promoted as a land use strategy which is capable of providing additional products and income in addition to environmental and conservation benefits. Such systems create a wide range of biophysical interactions between the associated plants. These interactions may consist of competition for water, light and mineral elements resources (Cannell et al., 1998; Bellow and Nair, 2003; Dufour et al., 2013). However, the different species can also create improvements in their shared environment: for example, a tree may improve the microclimate in its understory. This is known as 'facilitation' (Vandermeer, 1989). As plant species differ in their temporal and spatial patterns of resource acquisition, an intercrop may capture the available resources of light, water and nutrients more completely than a single species (Yu et al., 2015). The deep roots of trees serve as a 'safety net' below the root zone of herbaceous crops (Allen et al., 2004) and trees in agroforestry systems are able to recycle soil nutrients which leach down through the herbaceous crop rooting zone, thereby reducing groundwater contamination and increasing nutrient use efficiency in the system (Rowe et al., 1999; Wolz et al., 2018). Deep rooted trees may also take up water resources which are not accessible to the crop, thus increasing resource use by the association (Fernandez et al., 2008). If the development of the two

http://dx.doi.org/10.19103/AS.2018.0041.14

associated species is not simultaneous, one may use the available light while the other is not present or is in a leafless stage. This increased efficiency of resources use is termed 'complementarity' (Gathumbi et al., 2002). The key to success in agroforestry systems lies in choosing the optimum combination of trees and crops for the exploitation of spatial and temporal complementarity or facilitation in resource use (Cannell et al., 1996; Descheemaeker et al., 2013).

Thus the plantation design (i.e. tree density, tree rows orientation, alley cropping width), the chosen trees and crop species, together with tree and soil management, are the levers which can be applied in interaction with pedoclimatic conditions.

Modern agroforestry in temperate areas generally utilises alley cropping systems, defined as the practice of planting annual crops between widely spaced rows of woody plants or trees (Van Lerberghe, 2017d). Before planting, a detailed analysis is necessary to evaluate the benefits expected from the introduction of trees. These benefits are ecosystem services (Pantera et al., 2018) and can be divided into four groups:

1 Environmental services (Quinkenstein et al., 2009): the trees protect the soil against erosion, introduce organic matter into the soil through leaf fall and the turnover of fine roots, store carbon in deep layers of the soil, improve water quality by intercepting lixiviated nitrogen (Allen et al., 2004; Andrianarisoa et al., 2016) or other inputs, improve vegetal and animal biodiversity (Torralba et al., 2016) and moderate the microclimate (Jose, 2009).
2 Economic outputs (Pantera et al., 2018): the exploitation of tree products including timber, fuelwood, fodder, honey and fruits increases productivity and diversifies the farm resources.
3 Patrimonial investment: trees can be capital for the next generation or for the farmer's retirement.
4 Cultural and societal services, including improved landscape aesthetics, enhanced countryside value, increased opportunities for employment and education and the maintenance of cultural heritage (Pantera et al., 2018) (Fig. 1a and b).

The farmer must also be aware of the constraints of introducing trees. The plantation and maintenance of trees requires time and training if good results are to be achieved. The climatic conditions and soil properties must also be carefully considered before choosing the tree species and undertaking planting.

In order to produce high-quality timber, it is necessary to plant suitable seedlings, to protect them appropriately and to carry out pruning and other maintenance operations.

**Figure 1** Agroforestry landscape (a) compared to monoculture (b).

## 2 Choice of tree species

### *2.1 One or several species?*

The plantation of several different species in one field reduces the economic and sanitary risks. As the wood market changes, one timber species with an initially high economic value may depreciate within a few years. Monocropping also tends to facilitate the spread of diseases with trees, as with other species. The association of species or varieties reduces the likelihood of spreading diseases or pests. It also improves biodiversity through increasing variety within auxiliaries habitat.

Complementarity between species in the use of resources is enabled, for instance, when a phenological lag exists between them. Thus wherever it is compatible with the plantation goals, a mixture of species, or at least of varieties, is to be preferred over a monospecific plantation.

### *2.2 Species choice*

The choice of tree species must take into account (1) the commercial priorities of the owner, (2) the site characteristics and (3) the intercrop requirements:

- The first-choice criterion is the objective of the project. If timber production is the priority, the different rates of growth among species can guide the choice. For instance, if the farmer wants fast-maturing trees, poplars (Populus sp.) will be a good choice, with harvest around 20 years after plantation whereas sorb trees (Sorbus domestica L.) will require 100 years of growth before harvesting.
- Not every tree will be suitable for a given area (Fig. 2). A biophysical assessment is essential before choosing the species. Each tree species has specific requirements with regard to soil, climate and topography. Table 1 shows the requirements of the main tree species likely to be used in temperate alley cropping systems. Many soil characteristics have an

**Figure 2** Sunburn on the trunk of a young wild cherry tree.

influence on the development of tree roots: depth, texture, amount of stones, structure, pH, waterlogging, nutrients and organic matter content. A soil analysis can give valuable information on these factors. Temperature and rainfall are the two critical climatic variables which mostly impact tree survival and growth. The resistance of trees to drought is also linked with the water retention capacity and depth of the soil. The topography of the field will also significantly modify the effect of the local climate on the soil surface properties. For example, a steep slope will increase water run-off, thus decreasing water penetration in the soil; a south-facing field is often warmer and dryer than a north-facing one. Complementarity between trees and crops is the key to success. Many combinations are possible but if the associated plants (i.e. crop(s) and tree species) have different phenological cycles, depth of root systems and specific pests, the negative interactions will be reduced and complementarity or facilitation can exist. For example, the entire vegetative growth of winter cereals occurs during the period when deciduous trees have shed their leaves, causing a very low shade. Winter cereals also cause a deeper development of the tree roots as they dry the upper soil layer before tree budbreak (Fig. 3).

Guides are available to help match tree species to the edaphic and climatic conditions (Ellis et al., 2005; Orwa et al., 2015).

**Table 1** Main tree species usable for temperate agroforestry plantations and their characteristics

| Species | Soil properties | Climate | Qualities | Harvest year |
|---|---|---|---|---|
| *Sorbus torminalis* | Not shallow nor dry soils; pH = 5.5–8; clay soils, even calcareous, bears temporary waterlogging | Oceanic or continental; poor resistance to cold; not sensitive to late frost; good resistance to summer drought | Hardwood, brown to red; high value; used for cabinet making, veneer | 50–70 |
| *Alnus glutinosa* | Deep; pH = 5–7; wet clay soils | Temperate, not in Mediterranean area; not sensitive to late frost | Softwood, pink; used to make objects and guitars, paper and energy; the tree fix nitrogen; melliferous | 15–25 |
| *Alnus cordata* | Not shallow nor dry soils; pH = 5–6; light soils | Mediterranean; thrives in dryer soils than other alders | Softwood, reddish-orange; used to make objects; the tree fix nitrogen | 15–25 |
| *Castanea sativa* | Deep light and humid soils; pH = 5–7; intolerant of lime | Oceanic, Western continental, Mediterranean; the year-growth is sensitive to late spring and early autumn frosts | Mid-hardwood, light brown; used to make furniture, barrels and roof beams, posts; nuts, fodder, coppicing; melliferous | 20 years for nuts, 50 years for wood exploitation |
| *Quercus suber* | Deep light dry soils; pH = 5–7; intolerant of lime | Mediterranean, Western continental; thick, insulating bark that makes it well adapted to forest fires | Bark used to make cork plugs, fodder, heating; hardwood, yellow to brown, with very good mechanical properties, but difficult to work with. Used to make furniture, barrels, floor and roof beams | 90–120 |
| *Quercus robur* | Deep and humid soils, prefers fertile soil; tolerates flooding; pH = 5.5–6.5 | Northern oceanic, continental; poor resistance to cold and late frosts | Hardwood, yellow to brown, with very good mechanical properties, but difficult to work with. Used to make furniture, barrels, floor and roof beams | 60–100 |
| *Quercus pubescens* | Adapted to dry shallow calcareous soils; pH = 6.5–8 | Mediterranean; sensitive to late frosts | Hardwood, yellow to brown, with very good mechanical properties, but difficult to work with. Used to make furniture, barrels, floor and roof beams | 120–200 |
| *Quercus rubra* | Deep, light and humid soils; pH = 4.5–6.5 | Oceanic and mountain, eventually continental | Hardwood, brown to reddish, with very good mechanical properties, easy to work with; used to make furniture, floor and roof beams; acorns for animals | 45–70 |

(*Continued*)

**Table 1** (*Continued*)

| Species | Soil properties | Climate | Qualities | Harvest year |
|---|---|---|---|---|
| *Quercus petraea* | All type of soil texture, but not too dry nor too shallow; pH = 5–7 | Usual in continental area, can be in oceanic climate; poor resistance to cold and late frosts | Hardwood, yellow to brown, very good mechanical properties, but difficult to work with; used to make furniture, barrels, floor and roof beams, fodder (leaves and acorns) | 60–100 |
| *Sorbus domestica* | Light soils, without clay; bears dry and shallow soils; pH = 6.5–8; adapted to very calcareous soils | Oceanic and continental, found in Mediterranean area; barely tolerant to gold, but good resistance to late frosts | Very hard coloured wood with very good mechanical properties; used to make furniture, one of the most expensive wood in Europe; edible fruits; melliferous | 100–120 |
| *Cupressus sempervirens* | Bears dry, calcareous and shallow soils; pH = 6–8 | Mediterranean, found in oceanic area; sensitive to heavy frosts | Very hard pinkish, rot-resistant wood; used to make furniture and roof beams; production of essential oil | 90–100 |
| *Acer campestre* | Compact soils with good clay-humus balance; not on dry and shallow soils; pH = 5.5–8 | Oceanic and continental | White to yellowish mid-hardwood, with good mechanical properties, but not stable; used to make tools; melliferous | 50–60 |
| *Acer platanoides* | Soils with good clay-humus balance; not on compact nor too light soils; bears shallow and dry soils; pH = 5.5–8 | Eastern continental and mountain; very good resistance to late frosts | White hardwood, with good mechanical properties; Used for wood turning, lutherie, veneer, posts; melliferous | 60–80 |
| *Acer pseudoplatanus* | Soils with good clay-humus balance; bears compact, but not light and shallow soils; pH = 6–8 | Oceanic, continental and mountain; good tolerance to cold and late frosts | White to yellowish mid-hardwood, with good mechanical properties, used for wood turning, lutherie, veneer, posts; melliferous | 40–60 |
| *Gleditsia triacanthos* | Light soils; not compact, but bears shallow soils; pH = 5–8 | Mediterranean and Western oceanic; good tolerance to cold | Brown to reddish hardwood; Used to make furniture, crossbars posts and roof beams; fodder (pods), melliferous, dyeing; thorny tree, thornless forms exist | 60–90 |

| | | | | |
|---|---|---|---|---|
| *Fraxinus excelsior* | Adapted to light and compact, but not dry nor shallow soils; pH = 5–8 | Oceanic, continental and mountain; poor resistance to late frosts | White to yellowish mid-hardwood, with good mechanical properties but difficult to work with; used to make furniture, tools, objects and barrels; fodder (branches) | 40–65 |
| *Larix decidua* | Light soils but bears compact and dry, but not shallow soils; pH = 6–7 | Mountain and continental, found in oceanic and Mediterranean areas; very good resistance to cold, not to late frosts | Brown to red softwood, with good mechanical properties; used to make roof beams, posts, objects | 40–60 |
| *Prunus avium* | Light deep soils, intolerant to compact, dry and shallow soils and to waterlogging; pH = 5–8 | Oceanic and continental, found in well-exposed mountain and in Mediterranean area. Bad tolerance to cold, but good to late frosts; susceptible to sunburn on the trunk | Reddish mid-hardwood, with good mechanical resistance easy to work with; used to make furniture, carving; edible fruits; melliferous | 40–65 |
| *Juglans regia* | Light deep soils, intolerant to compact, dry and shallow soils; pH = 6–8 | Oceanic, Western continental and Mediterranean, susceptible to heavy frosts | Light colour to grey half-hardwood, easy to work with, not very resistant; used for veener, wood turning, furniture, luxury objects; edible fruits (nuts) | 60–80 |
| *Juglans nigra* | Light deep soils, intolerant to compact, dry and shallow soils; pH = 6–7.5 | Continental and oceanic, found in Mediterranean area, susceptible to late frosts | Half-hardwood, purplish, darker than the *regia* wood, easy to work with, used for furniture, veener | 60–80 |
| Hybrid: *Juglans nigra X regia or Juglans major X regia* | Light deep soils, intolerant to compact, dry and shallow soils; pH = 6–7.5 | Oceanic, continental and Mediterranean, susceptible to cold and late frosts | Half-hardwood, darker than the *regia* wood, easy to work, rather resistant; used for veener, wood turning, furniture. High phenotypical variability between trees | 40–50 |
| *Olea europaea* | Light soils, adapted to dry and shallow soils, but not compact nor too wet; pH = 6–8 | Mediterranean, susceptible to cold and late frosts | Brown, marbled, very hardwood; with good mechanical properties, very stable and easy to work with; used for wood turning, marquetry, tools and objects; edible fruits, after treatment | 90–110 |

(*Continued*)

**Table 1** (*Continued*)

| Species | Soil properties | Climate | Qualities | Harvest year |
|---|---|---|---|---|
| *Ulmus campestris* | Deep and fresh soils; pH = 7–8 | Oceanic, continental and Mediterranean, susceptible to cold but not to late frosts | Brown reddish hardwood, very easy to use; used for veener, furniture, roof beams; very susceptible to Dutch elm disease | 40–60 |
| *Populus* sp. | Light deep, humid soils (alluvial); intolerant to compact, dry and shallow soils; pH = 5–8 | Continental and oceanic, found in Mediterranean area, resistant to cold and late frosts | White to very light colour, very softwood; used for packaging, paper, matches, coppice, energy; can be used to dry humid areas | 15–25 |
| *Pirus* sp. | Light deep, humid soils; intolerant to compact, dry and shallow soils; pH = 5–8 | Continental and oceanic, found in Mediterranean area, poor tolerance to cold and late frosts | Pinkish to brown hardwood, rather easy to work with; highly sought for marquetry, furniture, lutherie, toys, objects; melliferous, difficult to prune because of its thorns | 90–110 |
| *Malus* sp. | Light deep, humid soils; intolerant to compact, dry and shallow soils; pH = 5–8 | Oceanic and continental, found at high altitude, resistant to cold and late frosts | Light reddish coloured hardwood, easy to work with; used for marquetry, wood turning, carving, fodder, melliferous, fruits | 80–100 |
| *Robinia pseudoacacia* | All kinds of soils; pH = 6.5–7.5 | Oceanic, continental and Mediterranean, resistant to cold and late frosts | Yellow to ochre hardwood, very good mechanical properties, rot-resistant, but difficult to work with; used to make tools, posts, garden furniture; fodder, melliferous; N-fixing tree; difficult to prune because of its thorns | 50–60 |
| *Tilia cordata* | Deep and humid soils, intolerant to dry soils; pH = 7–8 | Continental and oceanic, found in Mediterranean area, poor tolerance to cold but good resistance to late frosts | White to pale yellow softwood, very easy to work with; used for wood turning and carving; flowers for infusion, melliferous | 60–80 |
| *Liriodendron tulipifera* | Light deep, humid soils; tolerant to compact but not dry nor shallow soils; pH = 7–8 | Oceanic, mountain and continental, very resistant to frost, but intolerant to frosts when it is young | Pale yellow to greenish softwood, very easy to work with; used for veneer, furniture, lutherie | 70–80 |

**Figure 3** Association of pea and hybrid walnut in April. The pea begins flowering while the tree budburst is not beginning.

# 3 Tree planting

## 3.1 Planting pattern

From a landscape point of view, the line layout of the plantation may appear dull. This may be improved by irregularly spaced planting on the tree rows, by staggered repartition between the rows or by mixing tree species.

### 3.1.1 Tree density

The choice of tree density depends on the desired balance between trees and crops. The higher the tree density, the stronger the competition with crops will be. Research suggests that agroforestry systems are profitable with 50–100 trees $ha^{-1}$. However, 100–150 trees $ha^{-1}$ is recommended in order to allow for thinning, only keeping the good trees (Van Lerberghe, 2017d).

A clear area should be maintained at the end of each line to allow machines to turn without damaging trees.

### 3.1.2 Tree rows orientation

Four criteria must be taken into account when deciding on orientation:

- The shape of the plot: a rectangular shape allows more trees to be planted. Where a plot is almost square, it is possible to choose the orientation (Fig. 4).
- Availability of light for the crop growing under the tree shade: a north-south orientation, if the plot shape makes it possible, allows homogenous

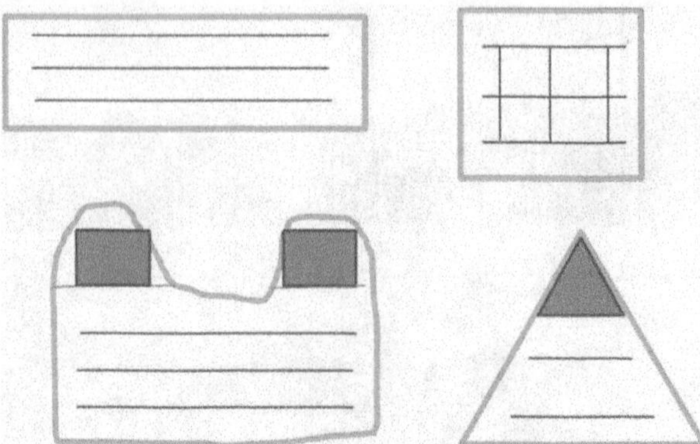

**Figure 4** Possible tree rows (brown lines) orientations according to the plot shape. The blue areas are lost surfaces for agroforestry, but can be wooded.

incident radiation on the crop. With this plantation scheme, each location of the alley cropping will receive light during the daily sun path. However, in an east-west-oriented field, the crop situated at the north of the tree lines will receive no light in the summer.

- Protection against the prevailing winds: in a windy area, the tree rows should be placed at right angles to the wind axis. A windbreak hedge can improve the protection effect.
- Anti-erosion efficiency: the vegetation of tree rows improves water infiltration, thus decreasing run-off. In sloping fields, planting tree rows along contour lines is recommended to facilitate the movement of agricultural machinery. In flooded areas, tree rows can be arranged across the current in order to slow it, to reduce soil erosion and to improve the fertile alluvium deposit. If the field is drained, the tree rows should not interfere with the drainage ditches.

### *3.1.3 Distance between the rows*

The distance between the rows determines the balance between crop and trees. If the intention is to maintain arable cropping through the whole tree rotation, the distance between the tree lines must be at least twice the height of the adult trees, that is, 30–40 m (Lawson et al., 2018). Narrower spacing, which gives increased timber production, is possible if the arable crop is replaced by grass lays in the last third of the rotation. The line spacing should also be compatible with the agricultural machinery used on the farm, allowing the use of the largest tools (spray booms for example) and being a multiple of the narrowest one (the seed drill for example). However, a problem may arise during the tree cycle where equipment changes over time. Figure 5 shows the example of a 13 m wide cropping alley which was designed for the equipment of 20 years ago but is no longer adapted to the modern combine width, so causing overlapping.

**Figure 5** Harvest in an agroforestry stand.

The non-cropped strip at the foot of the trees must be integrated into the calculation of the distance between tree lines. This strip is 1–4 m wide. A narrow strip requires great precision in the use of machinery, but gives a greater crop area and reduces the development of weeds which are often more competitive for young trees than the crop.

### 3.1.4 Distance between the trees on rows

The distance between the trees on rows generally varies between 4 m and 10 m. If it is <4 m, the competition between trees will be too strong and an early thinning will be necessary. If the distance is >10 m, the low density will not allow for choosing the best trees through thinning, and dead trees must be replaced to maintain a sufficient density (around 50 trees $ha^{-1}$).

## 3.2 Preparing the soil

The previous crop may affect the development of young trees though cultivated plots are generally easier to plant than unmanaged land. Where the soil of a cropped field was tilled in the previous year and the weeds were kept under control, no specific weeding is needed. In this case, ploughing the stubble is carried out after the crop is harvested. Weeding must be done before planting on grassland or fallow land to avoid competition which would cause a reduction of tree growth and possible mortality.

The weeding can be carried out by mechanical or chemical means; in the latter case, the remanence time must be taken into account. The false seedbed

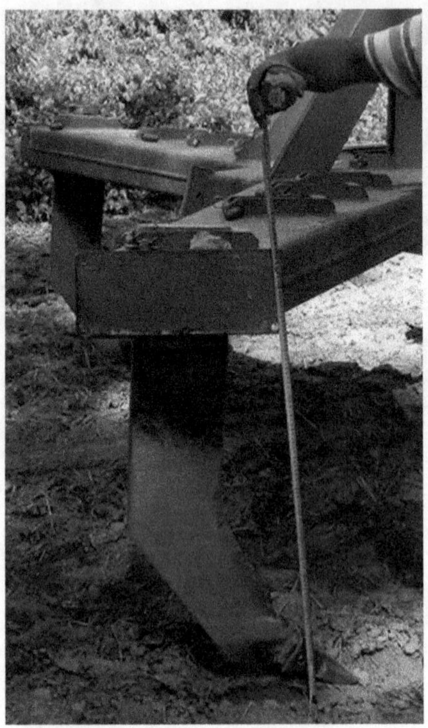

**Figure 6** Tool to relieve soil compaction to a depth of 80 cm.

technique may be used, which consists of working the soil at the end of summer as for normal sowing. This causes the seeds from previous crops and weeds to germinate. A second tillage three or four weeks later destroys the emerging plants, thus reducing the seed stock of annual weeds. However, this method is not efficient on perennials.

The soil must be prepared by subsoiling to a depth of >30 cm, especially if a plough pan is present (Fig. 6).

Finally, an additional shallow tillage or disking may be done on the tree plantation lines. This operation is necessary for mulching which requires a regular surface. A biodegradable or natural mulching (e.g. ramial chipped wood, straw, mowing residues) is preferable to a plastic film. Mulching improves seedling growth by reducing the development of weeds and limiting water loss through evaporation.

## *3.3 Picketing the plot*

Picketing is an important operation for marking the location of planted seedlings. The distance between tree lines must be regular and precisely

**Figure 7** Picketing the field with decameters and posts.

measured to avoid problems or damage during mechanical management of the crop. A GPS can be used for staking or it may be done with decameters and strings (Fig. 7).

### 3.4 Seedling quality

The key rule for a successful tree planting is to choose good quality planting material which will ensure good root development. As only a small number of trees are planted when compared to a forestry plantation, these must be of good quality. Seedlings from a specified region will be better adapted to local soil and climatic conditions.

#### 3.4.1 Genetic quality

Ideally, seedlings should be sourced from a certified supplier who will ensure the material is healthy and compliant with the expected genetic quality (Van Lerberghe, 2017a). The OECD Forest Seed and Plant Scheme defines four broad categories of forest reproductive material which are recognised for certification and identified by coloured tags:

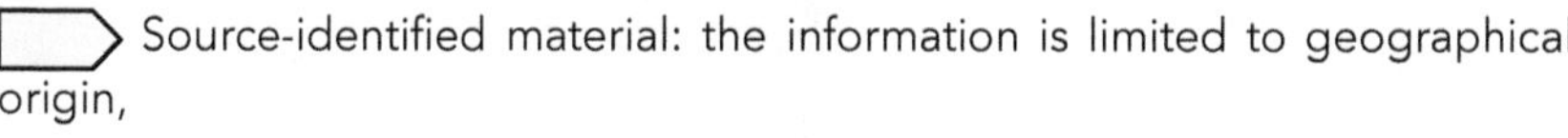

Source-identified material: the information is limited to geographical origin,

Selected material: harvested from selected stands in well-delimited regions of provenance,

Qualified material: the seeds are sourced from orchards where trees are individually selected according to vigour, shape, wood quality and disease resistance,

Tested material: harvested from stands or seed orchards where the superiority of particular characteristics has been proved through testing.

### 3.4.2 Types of plants

Four kinds of packaging are available for seedlings:

1 Bare-rooted, conditioned by size, in bunches 2-4 years after sowing. These seedlings are more fragile than the others, the advantage is that the sanitary condition and shape of the roots is visible, giving a good idea of a plant's state.
2 In single containers with an appropriate substrate. However, the root development may be limited by the shape of the container, so some nurseries offer seedlings in 'anti-bun' containers.
3 In clods: each seedling grown in a clump of earth without a container. This causes less disturbance to the root system but these seedlings are more expensive than those grown in a container. The transplantation stress is reduced for these two kinds of seedlings and when compared to bare-rooted plants, they are especially recommended for spring plantations.
4 Cuttings: for certain tree species (e.g. poplars and willows), unbranched shoots can be planted directly into the soil for rooting. In this case, there is no stress and the sanitary state is easy to check. Poplar cuttings should be strong and >1 m in length (Fig. 8a-d).

### 3.4.3 Age and dimensions

Nursery catalogues are precise as to the age (3 years maximum), the height and the collar diameter of the seedlings they sell. When possible, visiting the nursery prior to purchase gives a good idea of the seedlings quality and offers

**Figure 8** Different types of seedlings packaging: bare-rooted (a), container (b), clod (c) and cuttings (d).

an opportunity to seek expert advice. In selecting seedlings, it is advisable to choose plants as young as possible as these will have a high regeneration capacity, produce new roots quickly and are easier to handle than bigger plants.

### 3.4.4 Morphological quality

The vigour, architecture of the stem and roots if visible (for bare-rooted seedlings) are a guide to successful planting as well as to the sanitary state of the seedlings. Figure 9 shows a good stem shape (Van Lerberghe, 2017a).

The balance between root and stem volumes must be considered. The roots should be abundant, intact and have a good shape (see Fig. 10a–d).

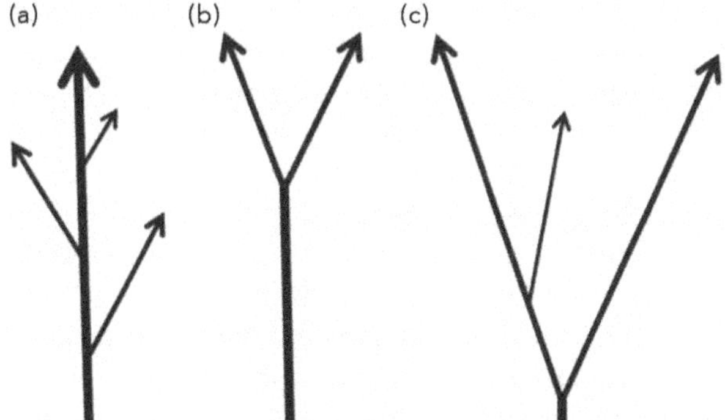


**Figure 9** Good shape of seedling stem and branches (a) vs split, with weak apical buds and frail stem (b and c).

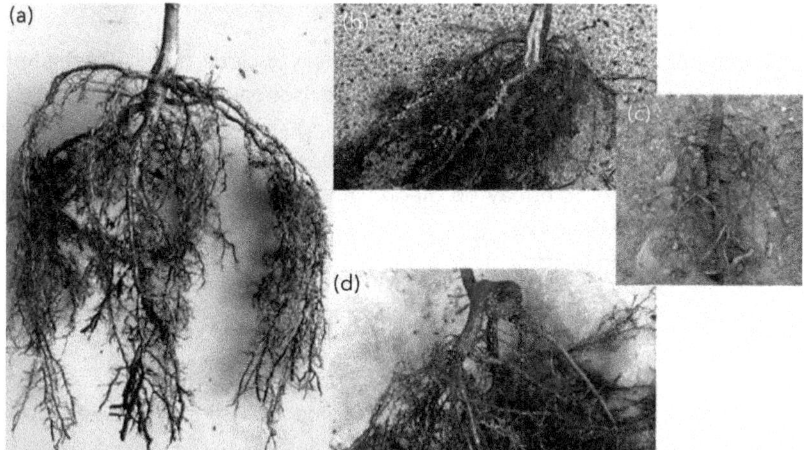


**Figure 10** Good root system shape (a), vs damaged collar (b), upgoing roots (c), hunting horn shape (d) (Van Lerberghe, 2017a).

### 3.5 Seedlings preparation

If the seedlings are to be planted more than two days after receipt, they must be stocked in a cool, shady place, sheltered from the wind. Bare-rooted plants must be kept in a trench, deep enough to take the roots and refilled with damp sand or soil. Cuttings should be placed vertically in a bucket of water.

Damaged or overlong roots must be cut off before planting. This helps the seedling to recover and should be done with sharp pruning shears. Any container must be removed before planting and the root hairs spread if necessary.

### 3.6 Planting

Bare-rooted seedlings and cuttings may be planted from November to March. Spring planting is preferable for heavy and humid soils, and autumn planting for light and sandy soils. Seedlings grown in containers or clods may be planted up to the end of April, but always before budburst. Cuttings are planted in the autumn. Snow, frost, wind or heavy rain are not favourable conditions for planting (Van Lerberghe, 2017c). The soil must be well prepared, neither too dry nor too wet, in order to avoid root rot.

If surface vegetation is still present, it should be removed within a 1 $m^2$ area around the planting areas. Square holes of 30 cm × 30 cm × 30 cm are dug with a spade. Bare-root seedlings must be settled in soft soil with enough space for the roots which should be well spread out in the planting hole. The roots can then be covered with fine earth.

To improve the contact of roots with the soil, place the collar around 2 cm below ground level and pull gently on the seedling stem after refilling the hole, so that the collar rises just above the soil surface. This improves the spread of the roots and avoids any risk of deformation which could hamper the growth and future stability of the tree.

When planting seedlings grown in containers or clods, the size of the hole should be approximately 10 cm larger than the root ball. The clods need to be immersed in water a few minutes before planting and re-covered with 2 cm of soil to prevent the substrate from drying out after planting (Fig. 11).

For cuttings, the hole may be made with a simple garden planter or an auger, according to their size (Fig. 12a and b).

Points to keep in mind:

- The stem must be vertical;
- The collar must be placed at ground level;
- The roots must be well spread in the planting hole;
- The earth covering the roots must be fine and slightly tamped down around the stem without damaging the collar;
- Watering is only necessary in very dry conditions.

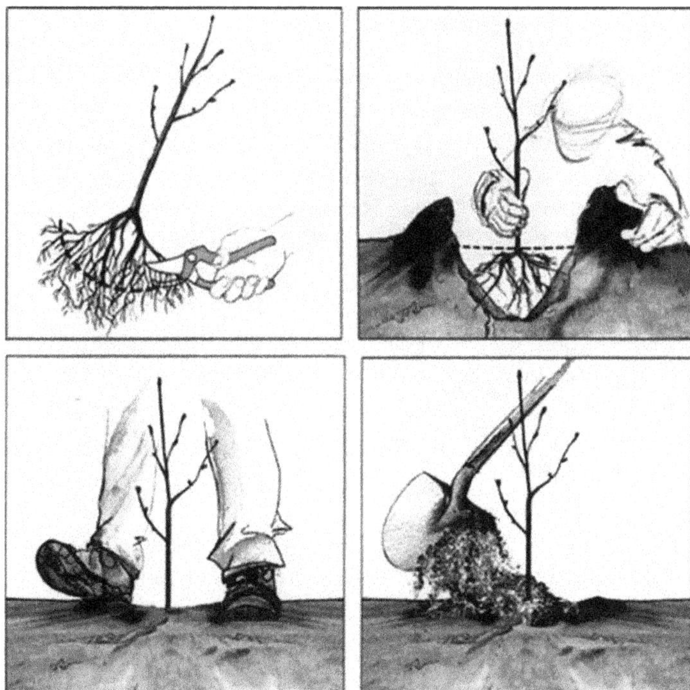

**Figure 11** How to plant a bare-root seedling (Dupraz and Liagre, 2008).

**Figure 12** Planting small (a) or big (b) poplar cuttings.

### 3.7 Young tree protection

Individual protection is essential in agroforestry systems because where there are only a few trees, they must all be preserved. There are a variety of guards with different properties. Mesh guard are less expensive, but also less safe than rigid tree shelters. All the devices need to be strengthened by a post that also supports the seedling. The height must be adapted to the risk: 0.5 m to protect against rabbits, 1.2 m for hares or only to segregate the tree from the crop, 1.7 m for sheep and roe deer and up to 2.2 m for cattle and deer.

Both ventilated and unventilated shelters have a greenhouse effect. They may be made of green, white or clear material and the ventilation system may consist of holes or slots. Ventilation at the bottom of the pipe (around 20 cm above the ground) ensures a regular renewal of the air inside it by the 'chimney effect' and provides the $CO_2$ necessary for young tree growth. These shelters have several functions:

- Protection of the tree stem against wild or domestic animals which can damage them by eating buds, leaves, branches or bark; by rubbing on the trunk. Deer may rub against tree stems to remove new antler velvet and rodents may gnaw bark at the collar or at the base of the trunk to wear down their incisors.
- Protection against contact with herbicides that may be sprayed on the crops or on the non-cropped strip.
- Identification of the tree location for the guiding of agricultural workers.
- Improvement of seedling growth, as a result of the greenhouse effect and the limitation of weed growth near the young tree.

Regular checking of the protection devices is important in order to straighten, repair or replace them where necessary. Checks must also be made to ensure that the protection does not hinder the tree growth or cause injury (Fig. 13a-d).

## 4 Plantation maintenance

### 4.1 Weeding

During the first years of tree growth, the presence of weeds near the seedlings must be avoided in order to reduce competition for water and nutrients.

#### 4.1.1 Mulching

The practice of mulching has been widely used as a management tool in many parts of the world. It reduces the influence of environmental factors on soil by increasing its temperature, controlling its diurnal/seasonal fluctuations (Ghosh

**Figure 13** Different kinds of seedlings protection: ventilated shelters (a and b), non-ventilated shelter (c) and mesh guard (d).

et al., 2006) and restricting weed development. If a bio-mulch is used, the thickness of the mulch layer should be maintained at a minimum of 10 cm and renewal is recommended at least every 2 years. The advantages of bio-mulch are: (1) it is wind-proof, (2) it reduces the overall cost, (3) it is environmentally friendly because it reduces chemical or mechanical weeding or replaces plastic mulch which is non- or only slowly biodegradable, (4) it reduces erosion, (5) it provides organic matter when decomposing, (6) it increases soil temperature and water content and (7) consequently, it improves tree growth.

The major disadvantage of this system is that mulching may increase or decrease the available N in the soil, depending on the C/N ratio of the mulching material. If the mulch has a high C/N ratio (>25 e.g. wheat straw), the immobilisation of N by microflora is likely to exceed mineralisation and the mineral N levels in the soil will decrease. The use of mature (at least 1-year-old) material can ameliorate this problem (Fig. 14).

### *4.1.2 Chemical weeding*

If there is no mulch, chemical weeding may be done in early spring in order to stop weed development as soon as possible. A second intervention may be necessary at the beginning of summer. The treatment must be applied on a day free from wind and rain, preferably early in the morning or in the evening, when the temperatures are not too high, and within a radius of 50 cm around

**Figure 14** Ramial chipped wood used as mulching.

each tree. The presence of non-perforated shelters will prevent seedlings from accidental contact with the product. Alternatively, a cover on the sprayer head can be used to direct the product stream away from the seedling.

### 4.1.3 Mechanical weeding

Mechanical weeding should be done manually to avoid possible damage to the seedlings by the use of tools. This is achievable where the tree density is low and the total plot area is not too large.

## 4.2 Non-crop strips management

Non-crop strips occupy a significant area of the field (3–10%), so there will be a decrease in production due to area loss. However, sub-tree vegetation can also host weed species or pests which may spread to the crop. This has to be balanced with the fact that the vegetation contributes significantly to biodiversity conservation in providing habitats and resources for pollinators and pest predators.

If the sub-tree vegetation is not systematically removed, some intrusive species, such as thistle, bramble or bindweed, will need to be controlled. Cutting the vegetation during spring with a brush cutter will prevent weed seeds spreading in the crop.

One method of limiting losses caused by the understory strips is to work the soil and plant the crop as close to the trees as possible. The tree roots are then confined under this surface and under the plough pan and will not be disturbed by successive soil management actions.

Another option is to cultivate the understory strips by sowing melliferous species which are attractive for some insects. Legumes may also be sown to enhance tree growth by fixing nitrogen and providing it to the soil. The understory strips can also be used to cultivate secondary crops such as berries and aromatic or medicinal plants.

## 4.3 Irrigation

Irrigating trees can be harmful if done frequently. This is especially the case with drop-by-drop irrigation which causes a limited and shallow root system to form, which will be undersized in relation to the aerial part of the tree. Regularly irrigated trees are very vulnerable.

However, supplemental watering may be necessary to ensure survival if there is drought during the 2 years after planting. It should be done mainly during the spring, using large amounts of water that will reach the deep layer of soil and make the roots grow downwards. Summer watering is only recommended where there may be a risk of the tree dying.

If the crop is irrigated, the trees will benefit from this water supply. However, it is unsafe to associate trees with irrigated crops over several consecutive years as this will modify the spatial distribution of the tree root system, and as a consequence, the trees will then have difficulty in surviving without watering.

Finally, the quality of the wood is affected by irrigation which causes fast growth without heartwood formation.

## 4.4 Fertilisation

In agroforestry systems, the tree does not require fertilisation as it exploits the fertiliser which is applied to the understory strips (if there is some) and their deep roots will capture the fertilisers that have leached below the crop root zone. Fertilisers may be applied to young trees during the first year after planting, provided an accurately calibrated boom sprayer that will not over-spray onto the non-cropped strips used.

# 5 Tree pruning and thinning

## 5.1 Tree pruning

### 5.1.1 General guidelines

The purpose of pruning is to improve the tree shape at its mature stage and to increase the production of high-quality wood without knots. Agroforestry trees are widely spaced in planting so they naturally develop a short trunk with many vigorous branches, while in forests, competition decreases the strength of the lower branches and branches in shade may die. Moreover, in an agroforestry plot, long branches may hamper the use of agricultural machinery.

The key reasons for pruning are:

- to improve the mechanical resistance of the tree and its ability to withstand wind,
- to preserve tree health by removing dead, broken or diseased branches,
- to lift the tree crown by cutting low branches to ease the farming operation,
- to provide a clean bole (the part of the trunk located between the soil and the first branches) of around 30–50% of the tree height and to allow the timber to be exploited as knot-free veneer logs or high-quality carpentry logs. This increases the value of the timber. The bole represents 90% of the tree value and must be straight and knotless. The knots result from the healing of pruning wounds. These cause deformations during drying and are problem areas for future use of the wood,
- to increase light penetration to crops.

However, it must be kept in mind that the branches carry leaves which feed the whole tree. Therefore, pruning must be gradual, frequent and precise (Chesney, 2012).

To prune a branch effectively, the cut must be clean, leaving no torn bark around it. It should be at a slight angle to the trunk and not too close in order to allow the underlying tissues to develop a healing callus. But if the cut is done too far from the trunk, it will leave a dead stump which will take several years to disappear, generating a high risk of infection.

Although dead branches can be removed at any time, it is recommended that trees should not be pruned during frost or periods of heavy sap flow. Living branches must never be cut during budburst, nor while the sap flow is going downwards (from the end of August to the leaf fall) in order to avoid depleting tree's resources. From late June, a moderate summer pruning of young trees is advisable. This summer pruning allows a good healing of wounds with a high resistance to pathogens and decreases epicormic growth. Pruning can also be undertaken in winter, when it is more easy to observe the underlying tree architecture. However, the epicormic growth is then likely to be more vigorous during the following spring.

Cut branches may be ground and used as mulch for the uncropped strip. They may also be exported as ramial chipped wood or as fuel.

### 5.1.2 Shape pruning

The purpose of shape pruning is to form the trunk axis. It consists of the removal of forks which divide the trunk in two axes, multiple, broken or defective top branches and those which are too big and compete with the main stem. The objective is to get the highest single, straight, cylindrical trunk possible, while giving the crown a shape that will allow the tree to withstand wind and ensure growth (Van Lerberghe, 2017b).

Pruning usually starts 2 years after planting or when the young trees appear out of the shelters. If a fork has formed in the shelter, it has to be cut by using hand pruning shears through the shelter or by lifting it. If there are shoots inside the shelter, these are removed at the same time. Then the shelter is re-positioned. It is not necessary to repeat this operation because the apical dominance of the upper part of the tree will prevent branches from growing inside the shelter. Weak seedlings which have difficulties in establishing or growing should not be pruned as this deprives them of a part of the foliage needed for development of the root system.

Shape pruning is carried out each year until the trunk is straight, without a fork and has reached the desired height. This will take around 10-25 years, depending on the tree growth. A moderate pruning every year is more efficient than a heavy pruning every 2 or 3 years.

Shape pruning is carried out from the top of the tree. Top branch defects are corrected as they represent a threat to the straightness of the trunk. Forks must be cut as soon as they appear. Lower vigorous branches are cut as soon as their diameter exceeds 3 cm. However, no more than 30% of living branches carrying photosynthetically active leaves must be removed annually. It is possible to cut larger branches (with a diameter between 3 cm and 6 cm) that were missed in previous years, but in order to obtain a clean wound and avoid ripping the bark this must be done in two successive steps: a first cut 30 cm away from the trunk followed by a second cut near the trunk. Branches with a diameter of 7 cm or above should not be cut as this could threaten the health of the tree and greatly reduce its growth (Fig. 15).

If a tree has a very bad shape or if the main stem is broken, it is wise to cut it back to around 5 cm from the ground before the beginning of spring. During the following summer, the most vigorous shoot growing from the stump should be selected to form a new trunk and the others removed.

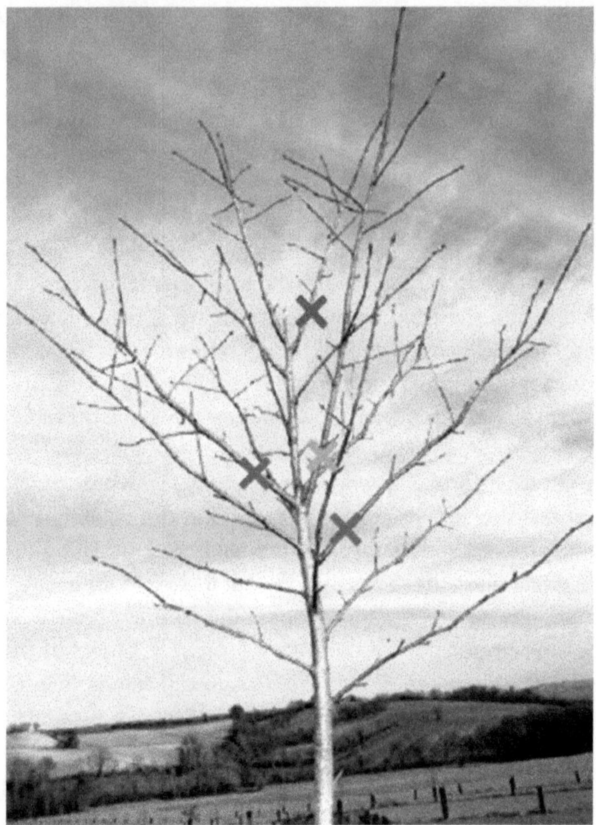

**Figure 15** Three operations for a good shape pruning: removing (1) the top fork (red), (2) the biggest branches (blue) and (3) the branches with an acute angle (green) (Van Lerberghe, 2017b).

### *5.1.3 Silvicultural pruning*

When a tree has been pruned into the desired shape, further pruning consists of the removal of lower branches to produce a quality bole, without defects, as high and cylindrical as possible and with the greatest proportion of knot-free wood. Knots should only be present at the core of the trunk, in a cylinder of 10–15 cm diameter, at the most.

Larger branches should not be cut if the grower wishes to avoid the formation of knots for this reason; silvicultural pruning should begin a few years after the shape pruning when trees have reached 3 m in height for walnuts and oaks and 4 m for other species. Trees must also be checked frequently, ideally each year, to deal with small branches of 3–4 cm in diameter at the most. These branches should not be cut before reaching 2 cm in diameter.

The smallest branches are left as these contribute to the tree growth, reducing the development of new shoots by acting as sap drawers. Pruning must be gradual to avoid unbalancing the tree, or the growth of epicormics. After the first pruning, the height of the branchless trunk should not exceed one-third of the total tree height.

Unlike shape pruning, silvicultural pruning is done from bottom to top, although hazardous large branches located at the top of the tree may be cut before lower ones with a smaller diameter. Branches are considered hazardous when they grow with an angle of <30° from the main stem (these may overtake the apical stem and create a fork) or when they reach a diameter equal to or greater than half that of the trunk.

Pruning facilitates the work of machinery in reducing the space occupied by branches, but branches of mature trees may spread out over the alley cropping. In this case, cutting can be undertaken with a mower bar. This is easy and quickly made and reduces the canopy without prejudice to the timber wood as the trunk is not affected. It also encourages the trunk to swell.

### 5.1.4 Specific case of pollards

Pollarding is a traditional practice which consists of periodically topping the trunk of a tree and maintaining steady pruning of all the branches (Chesney, 2012) without reducing the tree bole. This pruning system, which involves the removal of the upper branches of a tree, promotes a dense head of foliage and branches, producing a distinctive thick and bushy appearance.

#### 5.1.4.1 Productions generated by pollarding

The tree size and leaf area are significantly reduced during the first year after pollarding which historically was frequently practised for fodder or firewood production (Sjolund and Jump, 2013). For instance, pollarding ash trees produces 'pollard hay', which was traditionally used as livestock feed. Trees were pruned at intervals of 2–6 years in order to maximise leafy material.

Wood-producing pollards were pruned at longer intervals of 8–15 years, a pruning cycle that allowed the production of upright poles favoured for fence rails, posts and boat construction. Supple young willow or hazel branches could be harvested as material for weaving baskets, fences and garden constructions such as bowers.

Pollarding can be undertaken to combine biomass and timber wood production. The branches and leaves harvested at each pollarding may be used as fodder (ash tree), fire wood, ramial chipped wood or chips.

### 5.1.4.2 Pollarding practice

Many tree species can be pollarded to provide a range of products (Colin et al., 2017). Large pollards of fast-growing trees, such as willow or ash, will produce up to 30 kg of biomass per tree and per year.

Pollarding can be carried out every 5-15 years depending on the tree growth and biomass use. All the branches are removed when there is no sap rise, often in winter. However, pollarding may also be carried out in summer in order to provide fresh fodder during the period of grass shortage (Fig. 16).

### 5.1.4.3 Advantages and disadvantages of agroforestry pollarding

In an alley cropping system, pollarding may reduce the light and water competition between trees, and consequently may enhance crop yield (Dufour et al., 2016). Provided growth in the trunk diameter of timber trees is not too severely impacted, pollarding offers an effective way of obtaining three products simultaneously in an agroforestry field: annual crops, branch and leaf biomass for energy, fodder or wooden chips (Valipour et al., 2014) and timber wood. Although pollarding is an age-old practice that arose from the experience of farmers and foresters (Thomas, 2000), the growth and physiological responses of pollards have rarely been studied, particularly in the context of agroforestry.

The advantage of pollarding in an agroforestry system is the reduction of competition with the crop during the first years after pruning. However, especially in vigorous species, the trees recover their canopy quickly and a few years after pollarding, may produce a heavier shade than non-pollarded trees (Dufour et al., 2018).

**Figure 16** Pollarded ash trees.

### 5.1.4.4 An original experiment with pollards

An experiment was conducted to assess the impact of pollarding on the growth, both in trunk diameter and branch biomass, and development (phenology) of adult hybrid walnut trees, which are not traditionally pollarded, in an alley cropping agroforestry system (Fig. 17a–g).

The distance between tree rows is 13 m. In 2013, the walnut trees were 15 years old and their mean height was 10 m. In the same year, the crop yield was on average 25% lower in the agroforestry alleys than in the full crop control. Fifty trees were pollarded in December 2013 and their growth and development were compared to those of control trees, regularly pruned. The main results of this experiment were as follows:

- Pollarding changed the tree phenology: budburst was delayed during the first, third and fourth year after pollarding and the leaf fall was noticeably delayed for pollards when compared to control trees in the second year

**Figure 17** Pollarded walnut trees, at the first pollarding, in December 2013 (a), in December 2014 (b), in December 2015 (c), in December 2016 (d), in October 2017, just before (e) and after (f) the second pollarding, in May 2018, at the beginning of budburst (g).

after pollarding. As a result, the leaf-bearing time was extended by 26 days during 2015. This increase of leaf presence extends the photosynthetic assimilation of the tree and allows it to balance a part of the loss due to leaf area reduction.

- The pollards showed a significantly greater growth in height than control trees and except during the first year, their growth in diameter was not much slower than that of the control trees: cumulative growth from 2013 to 2017: 4.3 mm for pollards, 4.6 mm for control.
- The number of branches cut at the second pollarding, in October 2017, was 29.6 ± 0.4 (mean, standard error) per tree, with a mean branch diameter of 4.5 ± 0.07 cm. An allometric equation between the circumference of the branches and their dry biomass allowed an estimation of the biomass of branches produced between the two pollardings to be 80.9 ± 7.6 kg tree$^{-1}$, that is around 20 kg trees$^{-1}$ year$^{-1}$.
- The crop yield was only enhanced in the first 2 years because, as the photographs show, the trees quickly produce new vigorous branches and from the third year, the shade near the pollards is higher than that of the near regularly pruned trees.

Uncertainty remains as to how the trees will respond to repeated pollardings. The soil nutrients may be depleted due to the export of nutrient pools in the removed branches. This may increase the competition for nutrients with the crop. In addition, the rapid regrowth of branches creating very dense shade increases light competition with the crop as soon as the third year after pollarding. More studies are needed to optimise the trade-off between trunk (timber wood), branches (fuelwood or fodder) and crop (agricultural production) productivities, depending on the frequency of pollarding and the density of pollards on the field.

## 5.2 Tree thinning

Early elimination of poorly formed or weak trees which cannot be improved is important for successful agroforestry and the final density must allow the remaining trees to grow freely and without competition.

No real thinning is necessary when initial density is <50 trees ha$^{-1}$. For densities of 50–100 trees ha$^{-1}$, one tree out of two can be removed, and for higher densities, two out of three.

The thinning aims are: (1) to promote beautiful trees and to reduce the competition between trees, (2) to spend the working time only on crop trees and (3) to reduce the competition with the intercrop.

Thinning is undertaken as soon as possible to select the most beautiful trees as soon as the bole is formed. The bole is considered formed when forks have been removed and when the height of the tree is about twice that of the bole. There is no advantage in waiting for the bole to be totally pruned as the benefit of thinning to save pruning time will be lost.

The first thinning is carried out between 5 and 12 years after planting and if necessary, a second thinning may be done between 15 and 20 years after planting. After thinning, the final density is between 25 and 100 trees $ha^{-1}$. Thinning is selective, not systematic: several beautiful trees may be consecutive within a row, and they must be kept. Two or three trees can grow together even if their canopies touch. To achieve the most effective spacing, thinning may be undertaken in groups of three to six consecutive trees in the row. Half of the trees, the least beautiful in a group, will be removed.

Care must be taken when several tree species are planted, as growth rates differ greatly between species. It is important to take this characteristic into account and to keep trees of each species during the thinning.

The wood produced by early thinning has a low marketable value but may be ground or used as energy wood.

Many species will produce clump shoots after they are cut (e.g. walnut, poplar, maple, chestnut). The shoots can be used for biomass (ramial chipped wood, energy), otherwise the stump should be removed.

## 6 Conclusion and future trends

A decision to set an agroforestry plot requires an evaluation of the feasibility of the project in terms of land availability and fit, of skilled labour availability and of the correct tools. Success with an agroforestry plantation depends on the skill, experience and enthusiasm of the farmer who must also be knowledgeable in managing trees. These requirements could be met by two individuals, for instance the owner of the field takes care of the trees and a farmer manages the crops. This will require good coordination between the two roles.

Financial feasibility and return are key issues which farmers and land owners must consider. Regional differences in yields, prices and government grants might lead to substantial variations in revenue and costs within Europe and even within a given country. It is difficult to estimate the proceeds of tree sales due to the long period of time before harvesting, during which the selling price will fluctuate.

Graves et al. (2011) developed a new economic model: Farm-SAFE. This model allows comparison of arable farming, forestry and agroforestry systems at both plot- and farm-scale. The comparisons include net margins, net present values, infinite net present values, equivalent annual values and labour requirements. The new and free version of Farm-SAFE can be uploaded from: https://www.agforward.eu/index.php/fr/1683.html.

## 7 Where to look for further information

In many countries, associations of agroforestry can help the farmers to prepare and to realise their plantation project.

In the United States, the Association for Temperate Agroforestry (AFTA, https://www.aftaweb.org/), formed in 1991, promotes the wider adoption of agroforestry by landowners in temperate regions of North America. It is a private, non-profit, organization based at the University of Missouri Center for Agroforestry at Columbia. Also, the USDA developed a National Agroforestry Center (NAC). This agency accelerates the application of agroforestry through a national network of partners. They conduct research, develop technologies and tools, coordinate demonstrations and training, and provide useful information to natural resource professionals. For example, the centre published '*Profitable Farms and Woodlands: A Practical Guide to Agroforestry for Landowners, Farmers and Ranchers*' and the 'Agroforestry practices' section (https://www.fs.usda.gov/nac/practices/index.shtml).

In Europe, the EURopean Agroforestry Federation (EURAF, http://www.eurafagroforestry.eu/welcome) gives examples of agroforestry best practices. This association also reports on the place of agroforestry in the Common Agricultural Policy in the EU (http://www.eurafagroforestry.eu/action/policy/Agroforestry_in_the_new_CAP). EURAF sustained AGFORWARD project (AGroFORestry that Will Advance Rural Development), which was a 4-year research project funded by the European Union's Seventh Framework Programme for Research and Technological Development. One of the project outputs was the creation of two sets of leaflets on Best Practices and Innovations in agroforestry (https://www.agforward.eu/index.php/en/best-practices-leaflets.html and http://www.agforward.eu/index.php/en/Innovation-leaflets.html). On this topic, Rigueiro-Rodríguez, A., McAdam, J., Mosquera-Losada, M. R. published *Agroforestry in Europe: Current Status and Future Prospects* (2008; books.google.com).

Some books can also provide some guidance for the plantation and management of an agroforestry field: Dupraz, C. and Liagre, F. *Agroforesterie - Des arbres et des cultures* (book + DVD; Editions France Agricole, 2008) or for the tree species choice: Orwa, C., Mutua, A., Kindt, R., Jamnadass, R. and Simons, A. *Agroforestree Database: A Tree Reference and Selection Guide* (2015; http://www.worldagroforestry.org/af/treedb/). Finally, *Temperate Agroforestry Systems*, edited in 2018 by Gordon et al. brings together many examples of temperate agroforestry and will make valuable reading for all those working in this area as researchers, practitioners and policy makers.

## 8 References

Allen, S. C., Jose, S., Nair, P. K. R., Nkedi-Kizza, P. and Ramsey, C. L. (2004). Safety-net role of tree roots: Evidence from a pecan (*Carya illinoensis* K. Koch)-cotton (*Gossypium* hirsutum L.) alley cropping system in the southern United States. *For. Ecol. Manage.* 192, 395–407. doi:10.1016/j.foreco.2004.02.009.

Andrianarisoa, K. S., Dufour, L., Bienaime, S., Zeller, B. and Dupraz, C. (2016). The introduction of hybrid walnut trees (*Juglans nigraXregia* cv. NG23) into cropland reduces soil mineral N content in autumn in southern France. *Agrofor. Syst.* 90(2), 193–205. doi:10.1007/s10457-015-9845-3.

Bellow, J. G. and Nair, P. K. R. (2003). Comparing common methods for assessing understory light availability in shaded-perennial agroforestry systems. *Agric. For. Meteorol.* 114, 197–211. doi:10.1016/S0168-1923(02)00173-9.

Cannell, M. G. R., Van Noordwijk, M. and Ong, C. K. (1996). The central agroforestry hypothesis: The trees must acquire resources that the crop would not otherwise acquire. *Agrofor. Syst.* 34, 27–31. doi:10.1007/BF001229630.

Cannell, M. G. R., Mobbs, D. C. and Lawson, G. J. (1998). Complementarity of light and water use in tropical agroforests II. Modelled theoretical tree production and potential crop yield in arid to humid climates. *For. Ecol. Manage.* 102(2/3), 275–82. doi:10.1016/S0378-1127(97)00168-0.

Chesney, P. E. K. (2012). Shoot pruning and impact on functional equilibrium between shoots and roots in simultaneous agroforestry systems. In: M. L. Kaonga (Ed.), *Agroforestry for Biodiversity and Ecosystem Services – Science and Practice*. Croatia, www.intechopen.com. ISBN: 978-953-51-0493-3.

Colin, J., Van Lerberghe, P. and Balaguer, F. (2017). Farming with pollards – A productive way of pruning. *Agroforestry Innovation Leaflets*. Agforward Research Project. https://www.agforward.eu/index.php/en/best-practices-leaflets.html.

Descheemaeker, K., Bunting, S. W., Bindraban, P., Muthuri, C., Molden, D., Beveridge, M., van Brakel, M., Herrero, M., Clement, F., Boelee, E. and Jarvis, D. I. (2013). Increasing water productivity in agriculture. In: E. Boelee (Ed.), *Managing Water and Agroecosystems for Food Security*. Book Series: Comprehensive Assessment of Water Management in Agriculture Series. Vol. 10, pp. 104–23. doi:10.1079/9781780640884.0000.

Dufour, L., Metay, A., Talbot, G. and Dupraz, C. (2013). Assessing light competition for cereal production in temperate agroforestry systems using experimentation and crop modelling. *J. Agron. Crop Sci.* 199, 217–27. doi:10.1111/jac.12008.

Dufour, L., André, J., Le Bec, J. and Dupraz, C. (2016). Influence of tree pollarding on crop yield in a mediterranean agroforestry system. *3rd European Agroforestry Conference*. Montpellier, 23–25 May 2016.

Dufour, L., Gosme, M., Le Bec, J. and Dupraz, C. (2018). Impact of pollarding on growth and development of adult agroforestry walnut trees. *4th European Agroforestry Conference*, Nijmegen, 28–30 May 2018.

Dupraz, C. and Liagre, F. (2008). *Agroforesterie – Des arbres et des cultures*. Editions France Agricole, 413p. ISBN: 13-198-2-85557-150-8.

Ellis, E. A., Nair, P. K. R. and Jeswani, S. D. (2005). Development of a web-based application for agroforestry planting and tree selection. *Comput. Electron. Agric.* 49, 129–41. doi:10.1016/compag.2005.02.008.

Fernandez, M. E., Gyenge, J., Licata, J, Schlichter, T. and Bond, B. (2008). Belowground interactions for water between trees and grasses in a temperate semiarid agroforestry system. *Agrofor. Syst.* 74, 185–97. doi:10.1007/s10457-008-9119-4.

Gathumbi, S. M., Ndufa, J. K., Giller, K. E. and Cadish, G. (2002). Do species mixtures increase above- and belowground resource capture in woody and herbaceous tropical legumes? *Agron. J.* 94(3), 518–26. doi:10.2134/agronj2002.5180.

Ghosh, P. K., Dayal, D., Bandyopadhyay, K. K. and Mohanty, M. (2006). Evaluation of straw and polythene mulch for enhancing productivity of summer ground nut. *Field Crops Res.* 99, 76–86. doi:10.1016/j.fcr.206.03.004.

Graves, A. R., Burgess, P. J., Liagre, F., Terreaux, J. P., Borrel, T., Dupraz, C., Palma, J. and Herzog, F. (2011). Farm-SAFE: The process of developing a plot-and farm-scale model of arable, forestry, and silvoarable economics. *Agrofor. Syst.* 81(2), 93–108. doi:10.1007/s10457-010-9363-2.

Jose, S. (2009). Agroforestry for ecosystem services and environmental benefits: An overview. *Agrofor. Syst.* 76(1), 1–10. doi:10.1007/s10457-009-9229-7.

Lawson, G., Dupraz, C. and Watté, J. (2018). Can silvoarable systems maintain yield, resilience, and diversity in the face of changing environments? In: G. Lemaire, P. Carvalho, S. Kronberg and S. Recous (Eds), *Agroecosystem Diversity: Reconciling Contemporary Agriculture and Environmental Quality* (1st edn.). Academic Press, pp. 145–68. eBook ISBN: 9780128110515; Paperback ISBN: 9780128110508.

Orwa, C., Mutua, A., Kindt, R., Jamnadass, R., Simons, A. (2015). Agroforestree database: A tree reference and selection Guide Version 4.0. 2009. http://www.worldagroforestry.org/af/treedb/.

Pantera, A., Burgess, P. J., Mosquera Losada, R., Moreno, G., Lopez-Diaz, M. L., Corroyer, N., McAdam, J., Rosati, A., Papadopoulos, A. M., Graves, A., Rigueiro Rodriguez, A., Ferreiro-Dominguez, N., Fernández Lorenzo, J. L., Gonzáles-Hernández, M. P., Papanastasis, V. P., Mantzanas, K., Van Lerberghe, P. and Malignier, N. (2018). Agroforestry for high value tree systems in Europe. *Agrofor. Syst.* 92, 945–59. doi:10.1007/s10457-017-0181-7.

Quinkenstein, A., Wöllecke, J., Böhm, C., Grünewald, H., Freese, D., Schneider, B. U. and Hüttl, R. F. (2009). Ecological benefits of the alley cropping agroforestry system in sensitive regions of Europe. *Environ. Sci. Policy* 12, 1112–21. doi:10.1016/j.envsci.2009.08.008.

Rowe, E. C., Hairiah, K., Giller, K. E., van Noordwijk, M. and Cadish, G. (1999). Testing the safety-net role of hedge row tree roots by 15N placement at different soil depths. *Agrofor. Syst.* 43, 81–93. doi:10.1023/A:1022123020738.

Sjolund, M. J. and Jump, A. S. (2013). The benefits and hazards of exploiting vegetative regeneration for forest conservation management in a warming world. *Forestry* 86(5), 503–13. doi:10.1093/forestry/cpt030.

Thomas, P. (2000). *Trees: Their Natural History*. Cambridge University Press, Cambridge, UK, 286pp. doi:10.1017/CBO9780511790522.

Torralba, M., Fagerholm, N., Burgess, P. J., Moreno, G. and Plieninger, T. (2016). Do European agroforestry systems enhance biodiversity and ecosystem services? A meta-analysis. *Agric. Ecosyst. Environ.* 230, 150–61. doi:10.1016/j.agee.2016.06.002.

Valipour, A., Plieninger, T., Shakeri, Z., Ghazanfari, H., Namiranian, M. and Lexer, M. J. (2014). Traditional silvopastoral management and its effects on forest stand structure in northern Zagros, Iran. *For. Ecol. Manage.* 327, 221–30. doi:10.1016/j.foreco.2014.05.004.

Van Lerberghe, P. (2017a). Choosing quality planting material. *Agroforestry Best Practice Leaflets*. Agforward Research Project. https://www.agforward.eu/index.php/en/best-practices-leaflets.html.

Van Lerberghe, P. (2017b). Shaping the trees. *Agroforestry Best Practice Leaflets*. Agforward Research Project. https://www.agforward.eu/index.php/en/best-practices-leaflets.html.

Van Lerberghe, P. (2017c). Planting the trees. *Agroforestry Best Practice Leaflets*. Agforward Research Project. https://www.agforward.eu/index.php/en/best-practices-leaflets.html.

Van Lerberghe, P. (2017d). Planning an agroforestry project. *Agroforestry Best Practice Leaflets*. Agforward Research Project. https://www.agforward.eu/index.php/en/best-practices-leaflets.html.

Vandermeer, J. (1989). *The Ecology of Intercropping*. Cambridge University Press, Cambridge, UK, 237pp. ISBN: 0-521-34592-8.

Wolz, K. J., Branham, B. E. and DeLucia, E. H. (2018). Reduced nitrogen losses after conversion of row crop agriculture to alley cropping with mixed fruit and nut trees. *Agric. Ecosyst. Environ.* 258, 172–81. doi:10.1016/j.agee.2018.02.024.

Yu, Y., Stomph, T.-J. and Makowski, D. (2015). Temporal niche differentiation increases the land equivalent ratio of annual intercrops: A meta-analysis. *Field Crops Res.* 184, 133–44. doi:10.1016/j.fcr.2015.09.010.

# Chapter 3

## Temperate alley cropping systems

*Diomy S. Zamora, University of Minnesota, USA; Samuel C. Allen, New Mexico State University, USA; Kent G. Apostol, Independent Researcher and Editor, USA; Shibu Jose, University of Missouri, USA; and Gary Wyatt, University of Minnesota, USA*



### 1 Introduction

The intensive agricultural production systems in industrialized nations, such as the United States, have undergone modernization over the past century, which has subsequently led to a remarkable increase in crop yields. The implementation of modern agricultural practices has largely excluded trees from the rural landscape causing negative environmental impacts such as soil fertility depletion, soil erosion, non-point-source pollution and biodiversity losses. Recent studies, however, have shown that the integration of trees in the agricultural landscape using alley cropping may mitigate these negative impacts (Jose et al. 2000a,b, 2012; Thevathasan and Gordon 2004; Palma et al. 2007; Zamora et al. 2009a). Alley cropping, an agroforestry practice, involves the planting of row crops or forage in alleys formed by widely spaced single or multiple rows of trees or shrubs (Garrett and Buck 1997; Gillespie et al. 2000; University of Missouri Center for Agroforestry (UMCA) 2015). In temperate settings, the tree-row and alley arrangement allows the use of standard farm equipment and reduces the need for manual labour, in

http://dx.doi.org/10.19103/AS.2018.0041.04

contrast with the configurations of most alley cropping systems in the humid tropical regions.

Alley cropping - also known as silvoarable in Europe - has recently been recognized and adopted in temperate climates (Eichhorn et al. 2006; Regueiro et al. 2009), while having been a common practice in the tropics for many years. In North America, alley cropping is most popular in the southern and Midwestern United States (US) where high-value hardwoods are used to create alleys of various widths that support conventional row, forage or horticultural crops (Fig. 1). Such practices are developed with both production and conservation (protection) benefits. In this region of the US, cotton (*Gossypium hirsutum* L.), peanut (*Arachis hypogea* L.), maize (*Zea mays* L.), soybean (*Glycine max* L. (Merr)), wheat (*Triticum* spp.) and oats (*Avena* spp.) are important crops for alley cropping that are normally combined with trees such as black walnut

**Figure 1** Examples of temperate alley cropping systems: (a) black walnut intercropped with soybean; (b) pear intercropped with lettuce; (c) poplar intercropped with wheat; (d) pecan intercropped with forages; and (e) slash pine intercropped with cotton. Photo source: (a) and (e), University of Missouri Center for Agroforestry. (d), Shibu Jose, University of Missouri. (b) and (c), National Agroforestry Center, Lincoln, Nebraska.

(*Juglans nigra* L.), pecan (*Carya illinoinensis* K. Koch) (Allen et al. 2004a,b; Wanvestraut et al. 2004; Zamora et al. 2006, 2007) and some pine trees (*Pinus* spp.) (Zamora et al. 2009a,b; Hagan et al. 2009). However, the relatively large initial investment and long time needed to maturity for trees and shrubs is a substantial economic hurdle to adoption (Dyack et al. 1999), although leveraging multispecies systems (Malézieux et al. 2009) and high-value tree crops (Molnar et al. 2013) could lessen this burden, provided that farmers are aware of this option.

## 2 Potential of alley cropping

With 2.2 million agricultural farms in the US covering 373 million ha (USDA 2018), for example, alley cropping has the greatest potential to be implemented in temperate regions among the various agroforestry practices such as riparian buffers, windbreaks, forest farming and silvopasture. However, its adoption is potentially hampered by apprehensions of landowners, whose cultural practices do not typically employ tree-crop systems, and who are unfamiliar with the management skills of such complex systems (Zinkhan and Mercer 1997). In the tropics, alley cropping is often employed to address soil erosion issues, as farming often takes place in highly erodible areas.

While limited data exist that provide reliable estimates of the hectares established in alley cropping in the US, an estimate of land suitable for alley cropping is possible. In the Midwestern US, there are more than 8.1 million ha of non-federal croplands that are characterized as highly erodible soil, with an erodibility index (EI) $\geq 8$ (USDA NRCS 2007). The EI provides a numerical expression of the potential for a soil to erode considering its physical and chemical properties and the climatic conditions where it is located; the higher the index, the greater the investment needed to maintain the sustainability of the soil resource base if intensively cropped. Soils with a low EI (<5) are only slightly susceptible to erosion and farmers wishing to plant row crops have a wider range of alternatives on these lands. An EI from 8 to 15 indicates that cropland bare of cover could erode at least eight or more times the tolerance value while an EI of 15 or higher suggests that erosion could occur at 15 or more times the tolerance value (USDA NRCS 2007). Certain areas of the Midwest and other regions have soils that are typically susceptible to damage, and in many cases these soils are cultivated in a manner that leads to the loss of excessive quantities of topsoil. Thus, the potential for alley cropping on highly erodible cropland is significant, given that it can protect fragile soils through a network of fine roots and biomass produced by trees and companion crops.

Furthermore, there are many hectares of land that are under orchard management in the US, and could be adapted or converted to alley cropping practices especially during the establishment phase. Intercropping is a traditional practice that has been used in the establishment of fruit and nut trees for hundreds of years in both tropical and temperate areas. In the tropics, intercropping is commonly used to reforest marginal lands. In North America, intercropping with row crops has been shown to be feasible and a viable option for income diversification. For example, the use of a number of native tree species in pecan production has been examined in the Southeast (Gordon and Williams 1991). Similarly, approximately 7 million ha of pastureland have high potential for use as cropland and could be alley cropped. When alley cropping is properly planned and managed, it can provide landowners excellent market opportunities for their wood products (Smith et al. 2013). By selecting intercropping species for which known viable markets exist, landowners could improve the financial conditions of their farm operations. However, the success of alley cropping depends in many ways on its design.

## 3 Design considerations

Agroforestry, which is perceived as a practice that can provide optimum production while maintaining ecological sustainability, can enhance many of the biophysical cornerstones of ecologically sound agricultural production (Gordon et al. 1997). Temperate alley cropping has two distinct system components: trees or shrubs, and the companion crop(s). The design of alley cropping practice is flexible to reflect landowner needs and site potential. However, there are physical interactions (e.g. above ground competition for light, and below ground competition for water and nutrients) between the rows of woody species and the companion crop that should be understood and reflected in plans for alley cropping practice design (Jose et al. 2004). Proper tree/shrub companion crop selection, tree/shrub planting design, layout and orientations, and available farm area are among the major considerations of an adequate alley cropping design.

### *3.1 Tree or shrub selection*

The tree or shrub selection in alley cropping is the most critical aspect in designing an alley cropping system, as this determines the long-term sustainability and also profitability of the practice. Many tree species of high economic value exist that could be used in alley cropping. Species selection, however, must be based on matching the species both to the site and to the companion crop (Udawatta et al. 2005; Garrett et al. 2009; Jose et al. 2009; Jose 2011), which can be a challenging task. A tree or shrub species must be selected that best accommodates the sunlight requirements of a specific crop. The tree species should also satisfy the criteria established by the landowner, which would include providing income to the family, creating a microenvironment suitable for the companion crop, protecting the site and having wildlife benefits. Various species may be identified that are potentially capable of addressing these production and protection functions; however, these functions can only be fully achieved when species are compatible with the specific site's climate and soil conditions. Table 1 shows tree and shrub species suitable for alley cropping systems in a temperate region, particularly in North America.

**Table 1** Common tree and shrub species suitable for temperate alley cropping systems

| Species | Site requirements |
|---|---|
| Apple (*Malus* spp.) | Tolerate a wide range of soil types but prefer well-drained, sandy loam to sandy clay loams. Some rootstocks are more tolerant of heavier, wetter soils. An ideal pH would be around 6.5, but a pH of 5.5–7.5 is acceptable. |
| Black walnut (*Juglans nigra* L.) | Require fertile, deep soil with good drainage and pH of 6.0–7.5. Sensitive to spring frost injury, so avoid low-lying frost pockets if nuts are desired. |
| Pecan (*Carya illinoinensis* K. Koch) | Grow in full-sun environment, and prefer acidic, alkaline, loamy, moist, rich, sandy, silty loam, well-drained, wet and clay soils. |
| Chinese chestnut (*Castanea mollissima* Blume) | Require well-drained loamy to sandy loam soils as they are susceptible to root rot on wet, heavy soil; pH requirement of 5.5–6.5. |
| Hazelnut (*Corylus avellana* L.; *C. americana* L.) | Withstand wetter areas than chestnuts or walnuts, but still prefer well-drained soil with a pH of 5.5–7.5. |
| Aronia (*Aronia melanocarpa* Michx) | Tolerate a wide range of soil conditions but prefer well-drained soils with slightly acid to neutral pH. Full sun is required for optimal production. |
| Black currant (*Ribes* spp.) | Grow well in any soil but avoid heavy clay soils or standing water. Ideal pH is 5.5–6.5. Self-pollinating, but better crops occur with cross-pollination. |
| Saskatoon, juneberry, serviceberry (*Amelanchier* spp.) | Adaptable to a variety of soil types but do best on gently sloping hills with good air and soil drainage and a pH of 6.0–8.0. |
| American persimmon (*Diospyros virginiana* L.) | Adaptable to a variety of soils, from wet bottomland to thin ridge tops. Young trees can handle partial shade, but full sun is recommended for fruit production. |
| Honey locust (*Gleditsia triacanthos* L.) | Thrive in a wide range of soils and climate zones. They have wide adaptability and tolerance to drought, compaction and pollution. |
| Elderberry (*Sambucus canadensis* L.) | Adapted to many growing conditions, and grow well on well-drained loamy soil. Avoid planting in dry sites. |

Adapted and modified from Wilson (2007).

## 3.2 Tree planting design, layout and orientation

Plantings can consist of a single tree species or a number of species. Similarly, single tree rows or multiple rows may be used, depending on the objectives of the farmers.

### 3.2.1 Single species versus mixed species

The alley cropping planting design is flexible and can be done to achieve the landowner's conservation and production goals. The establishment of a pure orchard or tree plantation and converting it to an alley cropping may be best for some farmers; however, the mixing of tree species offers advantages to other farmers. The planting of two or more woody (tree) species in alley cropping with similar growth patterns and site requirements may provide greater economic and ecological diversity and create a more stable stand. There are factors to consider when deciding how many rows to establish and the arrangements of trees within the row depending on the desired benefits. These factors include (1) annual crop produced and the area removed from production by tree/shrubs rows; (2) desired tree/shrub crops and management needed to enhance production, including minimum widths needed for farm machinery operation; (3) erosion concerns that multiple rows and combinations of trees/shrubs/grasses can better address; and (4) wildlife habitats created through multiple rows of combined trees/shrubs/grasses (UMCA 2015).

### 3.2.2 Single- versus multiple-row sets of trees

Conventional wisdom might suggest that single-row tree patterns are better suited than multiple-row patterns for alley cropping. However, alley cropping is unique because unlike conventional forestry, two or more crops are grown in an intimate mixture on the same piece of land. Because of the nature of interactions (either facilitative or competitive) between system components (trees/shrubs and crops) and the goal of optimizing economic gain, a mix of trees (Table 1) and companion crops may be created to provide the highest economic potential to the landowner. The selection of a single or multiple rows of trees in planting design and practice depends on the ultimate product that the landowner will derive from the trees (Garrett et al. 2009). In some plantings, this may be single-row patterns; in others, it will be a multiple of rows. In the US, the selection of single- or multiple-row sets of trees is dictated in part by the federal government's cost-share programme(s) that a farmer may apply to in order to establish the alley cropping practice. It is noteworthy that some tree species do not adapt well to single-row configurations under the open conditions, especially when the ultimate goal is to have high-quality timber. Under such conditions, a triple-row configuration using a 'trainer' species in the outside rows that does not show a strong light response (e.g. pine trees) might be more economically viable.

Planting multiple rows consisting of more than one species or mixing species within a row must be carefully planned, as the juvenile growth patterns of different species can vary significantly. Multiple tree rows have, in some studies, also been shown to be better than single rows in optimizing wood and forage production, which suggests multiple rows might also be better with some other companion crops. Lewis et al. (1985) in their study involving assessment of the effects of single-row versus double-row configurations of slash pine (*Pinus elliotti* Engelm) on forage yield showed that the latter configurations were in several ways superior to single rows.

### 3.2.3 Between- (and within-) row spacing and tree orientation

The spacing between individual trees within rows and between rows, and the orientation of trees in the alleys, is a critical consideration in designing an alley cropping system, as this will directly impact aboveground competition for light. In meeting the needs of light-demanding species such as conventional row crops, spacing between tree rows can be wide while the within-row spacing is narrow. An East–West orientation of tree rows will maximize the sunlight received by an alley crop. Many alley crop varieties and species are shade intolerant (i.e. row crops, forages, small berry crops etc.), therefore, alleyways must be sufficiently wide enough to accommodate their light requirements. Choices related to between-row spacing are critical in alley cropping practices and will directly influence the system's success. Between-row spacing of trees depends on a number of management decisions including the emphasis placed on wood production versus some other tree-related crops (Garrett et al. 2009). For example, if the emphasis is on wood production, between-row spacing will be less than if nut production is emphasized. The spacing in alley cropping practice is also determined by the duration of the cropping regime selected. Studies in Missouri (Garrett and Harper 1999) and Florida (Wanvestraut et al. 2004) have shown that a single row spaced 12 m apart can likely only accommodate light and moisture needs of maize and cotton, respectively, within 10 years of establishment, when lateral roots are not pruned to minimize competition between tree and crops.

The spacing between trees in alley cropping practices will also vary depending on the environmental protection goals of the farmer. In the tropics, alley cropping practice is employed in many cases to control erosion problems, while optimizing the use of the land is of great importance among farmers in the temperate region. However, as earlier noted, severe soil erosion is also rampant in agricultural production systems in the US; thus, trees should be closely spaced to provide an immediate effect, or a shade-tolerant shrub with a shallow and highly diffused root system should be strategically combined with the trees.

### 3.3 Available farm machinery and operating requirements

Often, farmers in temperate regions are hesitant to adopt alley cropping because of farm equipment compatibility issues with the practice. Tree/shrub spacing should be adjusted based on multiples of the widest farm equipment to be used in the alley. Alley widths should be planned such that full or multiple passes of the equipment are utilized.

### 3.4 Selecting companion crops

Companion crops are planted in the alleys between tree rows. Corn, soybean, wheat, barley (*Hordeum vulgare* L.), oats, pumpkins (*Cucurbita* spp.), lettuce (*Lactuca sativa* L.), peas, among others have demonstrated success in alley cropping. The choice of companion crop with the trees will vary depending on the types of trees selected and the crop(s) desired by the farmer. There are three major groups of crops which can be grown in alley cropping. These include row/cereal and forage crops, fruits and specialty crops, and biomass-producing crops. Initially the growing environment in the alley will be favourable to row crops requiring full sun (corn, soybean, wheat) or forages. As trees grow taller and develop larger crowns and root systems, they will exert greater influence on the growing environment in the alley with increased shade, water and nutrient competition. While availability of sunlight is a major factor that determines how well row crops or forages perform in the alley, water and nutrient competition is even more significant. The tree canopy density will be partially determined by the spacing of trees within a row and the width between tree rows, and this spacing will also influence below ground competition.

## 4 Functions/benefits of alley cropping

The benefits of the alley cropping practice can be broadly classified as achieving financial productivity (i.e. production goals) and sustaining environmental integrity (i.e. protection goals).

### 4.1 Economic benefits

#### 4.1.1 Diversification of farm products and supplemental income

Alley cropping diversifies farm enterprises by providing short-term cash flow from annual crops while also providing medium- to long-term products from the woody components. Timber and non-timber products may contribute to income generation from the farm. Intercropping with row crops has been shown to be feasible in the US. The combination of trees and annual crops creates dynamic agroecosystems that when properly designed, may increase and diversify farm income. For example, in the US, studied systems have grown maize along with black walnut (Jose et al. 2000b; Gillespie et al. 2000), cotton along with pecan (Allen et al. 2004a,b; Wanvestraut et al. 2004; Zamora et al. 2008), and with slash pine (Zamora et al. 2009b), and ornamental shrubs with longleaf pine (*P. palustris*) (Hagan et al. 2009), while wheat is grown under rows of hybrid poplar (*Populus* spp.) in Ontario, Canada. Alley cropping also offers an opportunity to produce annual or perennial biomass crops for biofuel (Holzmueller and Jose 2012). In addition to the potential for producing nuts, berries and fruits, well-managed timber can provide a long-term investment. While opportunities abound in terms of crop diversification, it must be noted that caution must be observed when considering large-scale adoption of diverse cropping systems, as economic path dependency associated with intensive agricultural production (such as in the US Corn Belt) can work against the incorporation of diverse crops, as described by Roesch-McNally et al. (2018). They conclude that this issue needs to be further addressed by adjusting policy and economic incentives that foster diverse systems and reduce climate risk.

### 4.1.2 A source of food supply

The role of agroforestry in food security is now getting recognized. Food security entails having access to sufficient, safe and nutritious food to meet the dietary needs of an individual (Dawson et al. 2013). In the tropics (e.g. Philippines), the rapid decline in forest area can be attributed to the large and rapid conversion of the uplands into permanent annual cropping to meet the food requirements of a rapidly expanding population (Magcale-Macandog et al. 2010). This activity causes environmental problems such as accelerated soil erosion. However, implementing agroforestry such as alley cropping can address this problem. Planting fruit-bearing trees or shrubs integrated with annual crops (e.g. corn, soybean, wheat etc.) can serve as an immediate source of food for the farmer. Alley cropping is considered as a sustainable agricultural practice because it addresses the establishment of a resilient food system, which focuses on the pivotal role of enhancing food production in fighting food insecurity and achieving a certain degree of food supply.

The role of agroforestry, alley cropping in particular, in food security cannot be underestimated. Garrity (2004) outlined various food security issues that can be addressed by adopting agroforestry practices. These include eradication of hunger through the establishment of basic production systems in marginal lands based on agroforestry methods of soil fertility and land regeneration; poverty reduction among the rural poor through market-driven, locally led tree cultivation systems that generate income and build assets that increase purchasing power; and improvement in health and nutrition of rural communities from products of agroforestry systems. In general, food from trees in agroforestry systems is of particular importance to farmers in developing areas and may contribute up to 25–50% of their annual food requirements (FAO 2013).

## 4.2 Ecological benefits

### 4.2.1 Soil erosion control and water quality improvement

Soil erosion is a serious concern in agricultural production systems, particularly in areas with an EI greater than eight (UMCA 2015). Croplands lose soil at an average rate of 15 tons $ha^{-1}$ $year^{-1}$ with over 90% of the land losing at a rate above the sustainable level (Pimentel et al. 1995). Intensive agricultural production in highly erodible sites could result in erosion at a level above the tolerance (USDA NRCS 2007). Properly designed alley cropping practices can reduce erosion and sediment transport through the trapping capacity of the forested strip or in combination with ground covers. The practices can regulate erosion by developing a network of surface roots. Wanvestraut et al. (2004) and Zamora et al. (2007) in their separate assessment of root dynamics of pecan in alley cropping in Florida found that fibrous root systems of pecan concentrate in the upper 30 cm of the soil profile. These researchers concluded that such root concentration in the soil profile could play a major role in reducing soil erosion.

The excessive application of nitrogen (N) fertilizer in commercial agriculture and forestry can result in leaching of nitrate ($NO_3$-N) into surface and sub-surface drainage water. Ground water contamination through N leaching is a long-term environmental problem resulting from intensive agricultural production (Bonilla et al. 1999; Ng et al. 2000; Allen et al. 2004a,b; Jose et al. 2009). Over-application of N increases the production costs of farmers by millions of dollars every year (Marshall and Bennett 1998). The negative effects of nitrate leaching on rivers, lakes and residential wells are of increasing public and scientific concern. Thus, the effects of trees in alley cropping systems are of interest environmentally because the deep roots of trees may serve as a 'safety net' for capturing the N that is leached below the roots of the crops, and thus may help in improving nutrient use efficiency and in mitigating groundwater contamination (Rowe et al. 1999; Allen et al. 2004a,b; Zamora et al. 2009a). There is direct and indirect evidence of the safety-net hypothesis in the literature. A study in Sweden by Browaldh (1995) on the effects of tree harvesting on natural N levels in the soil showed that natural N levels increased in the vicinity of harvested sites due to the lack of N uptake by tree roots of harvested sites. In other studies, isotopic tracers such as $^{15}N$-enriched fertilizer have been used to trace movement of applied N in alley cropping systems (Rowe et al. 1999; Jose et al. 2000b; Allen et al. 2004a,b; Zamora et al. 2009a). For instance, Allen et al. (2004b) found that pecan tree roots played a significant role in alleviating groundwater nitrate leaching through their safety-net role in a pecan-based alley cropping system. Similarly, Zamora et al. (2009a) observed the same pattern in a loblolly pine alley cropping system intercropped with cotton.

### 4.2.2 Pest and disease protection

Exploiting the principles of biological control of crop pests is an important management strategy for farmers. Information on the population dynamics of insects, pests and their natural enemies is essential for utilizing habitat management as a tool for biological control of pests. Mixed systems such as alley cropping with diverse crops are known to encounter less pest problems than pure crops (Girma et al. 2000). Alley cropping systems differ widely in their plant diversity and are more diverse than monoculture crops. According to Stamps and Linit (1998), agroforestry in general is potentially useful for reducing pest problems because tree-crop combinations provide greater niche diversity and complexity than annual crops alone. This effect may be explained in one or more of the following ways: (1) wide spacing of host plants of the intercropping scheme may make the plants more difficult for herbivores to find; (2) one plant species may serve as a trap crop to prevent herbivores from finding the other crops; (3) one plant species may serve as a repellent to the pest; (4) one plant species may serve to disrupt the ability of the pest to efficiently attack its intended host; and (5) the intercropping situation may attract more predators and parasites than monocultures, thus reducing pest density through predation and parasitism (Root 1973; Vandermeer 1989; Jose et al. 2004, 2009).

The effects of multi-species systems in controlling pests and diseases have been demonstrated in many studies. Altieri (1983) found that intercropped maize in Colombia recorded 25% fewer leafhoppers (*Empoasca kraemeri*), 45% fewer beetles (*Diabrotica balteata*) and 25% less fall army worms (*Spodoptera frugiperda*) than sole-cropped maize. Studies with pecan, for example, have looked at the influence of ground covers on arthropod densities in tree-crop systems. Bugg et al. (1991) observed that cover crops (e.g. annual legumes and grasses) sustained lady beetles (Coleoptera: Coccinellidae) and other arthropods that may be useful for biological control of pests in pecan. Eckberg et al. (2015) also observed that alley cropping involving willow and soybean provided host conditions for aphid predators on soybean. Stamps et al. (2002) observed similar results in a black walnut-based alley cropping system in Missouri in the Central US. In another study of alley cropping trials with peas (*Pisum sativum* L.) and four tree species (*Juglans*, *Platanus*, *Fraxinus* and *Prunus* spp.), Peng et al. (1993) found an increase in insect diversity and improved natural enemy abundance compared to monoculture peas. The greater diversity of birds in alley cropping systems (Gillespie et al. 1995; Gibbs et al. 2016) could also provide beneficial services for pest reduction. In essence, the diversity of species in alley cropping systems reduces damage from insect pests by reducing crop visibility, diluting pest hosts due to plant diversity, interfering with pest movement and creating habitat more favourable to beneficial insects.

### 4.2.3 Wildlife habitat enhancement

Intensive agricultural production and continued fragmentation of forest cover has been shown to have detrimental effects on wildlife populations (Wilcove et al. 1998; MacLeod et al. 2012; Vasileiadis et al. 2013; Gibbs et al. 2016). Wildlife species living within agroecosystems, including avian species, are especially impacted by conventional monocrop agriculture, which provides negligible nesting, foraging and refuge habitat (MacLeod et al. 2012), and has been linked to biodiversity losses (Pekin and Pijanowski 2012). Many solutions aimed at mitigating the consequences of unsustainable agriculture and subsequent biodiversity loss are suggested in the literature. One such strategy is the implementation of agroforestry practices in the landscape as a means of creating more heterogeneous and complex ecosystem structures than are found in conventional agricultural systems (Bakermans et al. 2012; Gibbs et al. 2016).

Alley cropping can play a vital role in providing wildlife habitat and landscape improvement. Incorporating the practice into an agricultural landscape increases the habitat diversity for wildlife (Garrett et al. 2009). This increased diversity improves habitat conditions and can increase wildlife use. Several studies have shown the positive impact of linear plantings on diversity and density of bird species. Yahner (1982) found that increased vertical complexity of the woody vegetation is correlated with increased bird numbers, and must be used in designing alley cropping systems. Increased vertical complexity of managed grassland in North Dakota, USA, has also been shown to increase bird usage (Renken 1983). The corridors created in alley cropping could provide more woody cover and foster greater total numbers of birds and more bird species (Garrett et al. 2009; Gibbs et al. 2016). Another advantage of linear plantings such as those in alley cropping is found in their edge effects. Because of their extensive edge-to-volume ratios, alley cropping can serve as a reservoir for many kinds of insects that attract bird species. Strong correlations between the amount of edge and bird species and bird abundance have been observed (Best et al. 1990).

Alley cropping may be viewed as a complex series of tree-crop interactions guided by utilization of light, water, soil and nutrients. Therefore, an understanding of the biophysical processes and mechanisms involved in the utilization of these resources is essential for the development of ecologically sound agroforestry systems. The following section discusses key issues, limitations and challenges of implementing alley cropping systems.

## 5 Competition for growth resources

The incorporation of multiple species in single ecological systems such as alley cropping brings about a unique set of ecological interactions among different species (Jose et al. 2004). By combining multiple resources in the same system, potential exists to increase nutrient use efficiency, control sub-surface water levels, improve soil and water quality, provide favourable habitats for plants, insects, or animal species and create a more sustainable agricultural systems (Garret and McGraw 2009; Garrity 2004; Garret et al. 2009; Jose et al. 2009) as a result of their component interactions. According to Nair (1993), interaction refers to the influence that one or more components of a system has on the performance of both another component of the system and the overall system itself. If two species are competing for the same resources, and do so equally, both species will likely exhibit lower productivity when compared to their potential for independent growth. However, alley cropping systems can be designed so that the physiological needs for a particular resource are spatially or temporally different for the individual species growing in the system. As such, then the system would lead to a greater productivity than the cumulative production of individual system components when grown separately on equal land area. As an association of plant communities, alley cropping is deliberately designed to optimize the use of spatial, temporal and physical resources by maximizing positive interactions (facilitation) and minimizing the negative ones (competition) between trees and crops (Jose et al. 2000b; Zamora et al. 2006). These interactions are outlined below.

### *5.1 Above ground competition for light*

The productivity of alley cropping is dependent on the crop-tree interactions. Understanding of the biophysical processes and mechanisms of resource capture such as above ground competition for light in alley cropping is necessary. Light availability and its effect on above ground production are equally important to agricultural system sustainability. Trees or shrubs in alley cropping can increase the amount of shading that plant species, primarily those in the understorey, experience compared to growing in a monoculture. When plant growth and productivity are not limited by water or nutrients, they can be altered by the amount of radiant energy intercepted by the foliage (Monteith et al. 1991).

Numerous studies have examined shading and its effect on crop growth (Monteith et al. 1991; Gillespie et al. 2000; Reynolds et al. 2007), and those studies indicate that light availability caused by shading is a major factor influencing production (e.g. reduction in crop yield). Reynolds et al. (2007), in their study on alley cropping in Ontario, Canada, showed that low levels of photosynthetically active radiation (PAR), 400–700 nm wavelength, resulting from overhead shading by hybrid poplar (*Populus* spp.) and silver maple (*Acer saccharinum* L.) significantly reduced maize and soybeans yields. Similar results have also been reported in a temperate silvopastoral system, although pastures may be somewhat less prone to reduced productivity due to greater overall surface area (Pardini et al. 2008). Lin et al. (1999) observed a significant decrease in mean dry weight of warm season grasses under silvopasture in Missouri, USA. However, cotton production under pecan alley cropping system in Florida demonstrated that light availability was not a major factor influencing cotton yield (Zamora et al. 2008). See Section 6 'Evaluating system performance' for details of the impacts of shading on cotton production.

The physiological response of plants to shading depends on the pathway used by crops to fix carbon ($C_3$ or $C_4$). In $C_3$ plants, as PAR increases from complete shade to approximately 25–50% of full sun, there is a corresponding increase in photosynthetic rate ($P_{net}$) (Fig. 2); however, as more light becomes available, $P_{net}$ does not increase, but rather levels off despite the additional increase in PAR. In contrast, in $C_4$ plants, $P_{net}$ does not level off as PAR increases in full sunlight, but rather continues to increase with increasing PAR (Fig. 2). In accordance with the theory that $C_3$ plants would not have reduced growth growing under shaded conditions, Zamora et al.

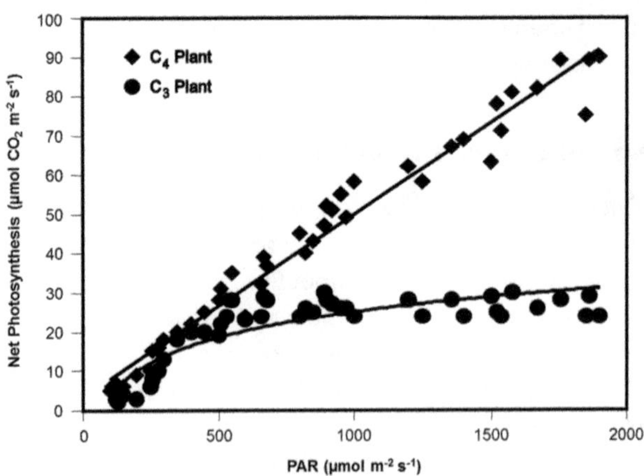


**Figure 2** Net photosynthesis of $C_3$ (cotton) and $C_4$ (maize) plants in response to PAR growing in a pecan and black walnut alley cropping, respectively in Florida and Indiana. Source: Adapted and modified from Zamora et al. (2008) and Jose et al. (2009).

(2008) concluded that light was not a major factor influencing production of cotton, a $C_3$ plant species growing under moderate shading in a temperate pecan alley cropping system described above in Florida, USA. Contrary to an anticipated yield decrease in maize, a $C_4$ species, in response to shading, Gillespie et al. (2000) reported that there are no effects of shading in both a black walnut - maize and red oak (*Quercus rubra* L.) - and maize alley cropping system in Indiana, USA, which was not expected given the known strong positive correlation between PAR and $P_{net}$ in $C_4$ plants. Although Gillespie et al. (2000) found that the edge rows received lower PAR compared to the middle rows, particularly in the red oak-maize system because of the higher canopy leaf area, once competition for water and nutrients was removed through trenching and polyethylene root barriers, there was no indication of yield reduction because of reduced PAR, leading the authors to conclude that light was not a limiting factor of these systems, but results from other studies may differ on this point.

## 5.2 Below ground competition for water and nutrients

Minimizing resource competition between trees and crops while maximizing the use of available resources is central to improving yield and overall productivity in alley cropping and similar agroforestry systems. The component species in alley cropping systems share the same biophysical resources and draw from a common resource pool.

### 5.2.1 Competition for water

Researchers in the temperate zone, humid tropics and semi-arid tropics alike have identified competition for water as the major limiting factor influencing productivity in alley cropping systems (e.g. Lehmann et al. 1998; Govindajaran et al. 1996; Jose et al. 2000b; Wanvestraut et al. 2004). In other studies, although certain alley cropping configurations generated benefits, they were overshadowed by production losses due to competition for water. For example, Govindajaran et al. (1996) in an alley cropping stand of maize and leucaena (*Leucaena leucocephala* Lam) found that nitrogen requirement for maize was met through the presence of leucaena but maize yield decreased 39-49% due to water competition. In a temperate alley cropping system involving hardwoods (oak and black walnut) and maize, Gillespie et al. (2000) reported crop yield reductions of up to 40% due to severe competition for water when trees were 11 years old.

Drought stress has been found to affect crop plants. Multiple studies have reported large reductions in crop plant height (NeSmith and Ritchie 1992; Wanvestraut et al. 2004) and leaf area (NeSmith and Ritchie 1992; Jose et al.

2000b; Wanvestraut et al. 2004; Zamora et al. 2008), and yield (Gillespie et al. 2000; Wanventraut et al. 2004; Zamora et al. 2007) when water availability is limited. In a silvopastoral study in the Canterbury region of New Zealand, Yunusa et al. (2005) reported a slightly lower water potential in radiata pine (*Pinus radiata* D. Don) trees planted in an alfalfa (*Medicago sativa* L.) pasture compared to trees growing in a vegetation-free environment. In that study, competition for water led to a decrease in photosynthetic rates in radiata pine compared to trees growing in the alfalfa-free treatment.

Competition for water in alley cropping systems becomes increasingly intense when water availability decreases throughout the soil profile (Miller and Pallardy 2001; Jose et al. 2009). Factors that limit plant growth and productivity due to competition for water include drought, soil water-holding capacity and irrigation. Competition for water in alley cropping can be minimal given adequate levels of either or both precipitation and irrigation. Furthermore, competition for water can also be minimized by proper tree-crop selection and combinations, as some deep-rooted trees have hydraulic lift capabilities. Hydraulic lift is the process in which some deep-rooted plants take in water from lower soil layers and release that water into upper, drier soil layers. This phenomenon has been reported to be an appreciable water source for neighbouring plants (Corak et al. 1987; Caldwell and Richards 1989). Trees in alley cropping systems can benefit associated crop plants through the mechanism of hydraulic lift, thus providing more water for surrounding vegetation during dry periods (van Noordwijk et al. 1996; Burgess et al. 1998; Ong et al. 1999; Jose et al. 2009). Recent evidence shows that this phenomenon can help promote greater plant growth, and could have important implications for net primary productivity, as well as ecosystem nutrient cycling and water balance (Horton and Hart 1998).

### 5.2.2 Competition for nutrients

Below ground competition for nutrients is most likely to occur when two or more species have developed a specialized root system that directs them to explore the same soil strata for resources (van Noordwijk et al. 1996). Thus, the extent of competition in a mixed system such as alley cropping will depend on nutrient availability, root architecture, rooting depth and proximity to competing roots, and temporal demands (Jose et al. 2000b; Allen et al. 2004a,b). As in the case of conventional agriculture, nutrients can often be limiting in agroforestry systems. Nitrogen (N) is generally the most limiting soil nutrient in temperate alley cropping systems due to various reasons. First, N is lost via various biogeochemical mechanisms such as volatilization, denitrification and leaching. N is also lost when crop biomass is removed from the field following harvest. In addition, plants of the same species and growth stage can compete heavily for N when zones of depletion in the soil overlap with neighbouring plants. In alley cropping systems, competition for nutrients can be even more intense, as most tree species have the bulk of their fine, feeder roots in the top 30 cm layer of the soil. The root systems of all component species in three temperate alley cropping systems consisting of maize and black walnut or red oak (Jose et al. 2000b), and pecan-cotton (Zamora et al. 2007) were found to most heavily occupy the top 30 cm of soil and to decrease in density with depth. Consequently, yield of maize was reduced by 35% and 33% in the black walnut and red oak systems (Jose et al. 2000b), respectively, and yield of cotton was reduced by 38% in the pecan system (Zamora et al. 2008).

Although it is difficult to separate the below ground competition for water from that for nutrients, it is widely recognized that crop production in agroforestry systems in semi-arid regions is likely limited by competition for water. When considering alley cropping systems in temperate regions, it is not clear as to which competitive vector will limit productivity, although competition for water has been reported (Jose et al. 2000b; Wanvestraut et al. 2004). Research on competition for nutrients in temperate agroforestry is rather limited (Jose et al. 2000b; Miller and Pallardy 2001; Allen et al. 2003; Zamora et al. 2009a). Many of the reported agroforestry systems were fertilized at the conventional agronomic level as needed for the crop component. Jose et al. (2000) reported that competition for nitrogen fertilizer was minimal in their black walnut-maize alley cropping system, since nutrient acquisition was not simultaneous among the system's components. Water availability, however, was a factor in nutrient competition, as competition for water by tree roots was responsible for the biomass reduction in intercropped maize resulting in decreased efficiency of fertilizer use (Jose et al. 2000b). The same trend was also reported by Zamora et al. (2009) in the production of cotton in a loblolly pine alley cropping system in Northwestern Florida. Under resource competition, plants tend to inhibit further growth. For example, in that study, cotton plants intercropped with loblolly pine showed reduced growth, and so were unable to utilize the available fertilizer efficiently. This phenomenon was also observed for maize growing in a black walnut system in Missouri (Jose et al. 2000b) and with cotton planted under pecan (Allen et al. 2004b) due to competition for water.

### *5.2.3 Allelopathy*

The mixture of multiple species in an alley cropping system may not only increase competition for growth resources such as water, light and nutrients but may bring about other negative effects such as allelopathy. Allelopathy is a biological phenomenon by which one plant may produce and release some chemical compounds that inhibit the growth of nearby plants. These chemicals are released into the soil by root exudation and aboveground litterfall.

Allelochemicals are known to affect germination, growth, development, distribution and reproduction of a number of plant species (Inderjit and Malik 2002; Jose and Gillespie 1998a,b; Jose et al. 2009). Several examples of allelopathy have been shown in temperate agroforestry systems (Jose and Gillespie 1998b; Thevathasan et al. 1998; Jose and Holzmueller 2008) such as those based on black walnut or pecan. These authors reported declines in yields of maize, cotton and peanut when these companion crops were exposed to juglone in hydroponic cultures. However, the effects of allelopathy can be minimized (at least in research plots) by separating the roots of the tree from the crop through the use of polyethylene barrier treatment (Jose and Gillespie 1998a,b). Jose et al. (2000) demonstrated that by using this barrier approach, maize crop yield became similar to that of monoculture maize. Jose and Gillespie (1998a,b) showed that the barrier also eliminated detectable levels of juglone in the soil in a black walnut alley cropping system. Perhaps the most effective approach to address allelopathic effects is proper species combinations when considering an alley cropping system as a production system.

## 5.3 Design complexities

While alley cropping is practised in temperate regions it is not widely acknowledged or adopted because of its apparent complexities. In their survey to identify landowners' and extension professionals' perceptions of agroforestry in the Southeast US, Workman et al. (2003) identified the lack of skills and machinery compatibility with the alley cropping operations as major factors in the slow adoption of practices. As with other forms of multi-cropping systems, alley cropping requires more intensive technical management skills and marketing knowledge. Alley cropping requires a more intensive management system including specialized equipment for the tree management and additional managerial skills and training to manage multiple crops at a given site. Removal of land from annual crop production is another barrier to alley cropping adoption because the practice may not provide a financial return from the trees for several years. Further, trees in alley cropping may be an obstacle to crop cultivation if not carefully planned and designed. Alley cropping also requires a marketing infrastructure for the tree products that may not be present in the local area. As earlier discussed, the competition for growth resources between the trees and crops in alley cropping systems may limit productivity; thus, proper design of the practice may address or minimize the impacts of these issues.

## 6 Evaluating system performance: the case of the pecan-cotton alley cropping system

Realization of the potential benefits of alley cropping depends on the production, protection (environmental benefits) and adoptability of the practice. To shed light on these three criteria (production, protection and adoptability), it is useful to examine them in the context of a specific documented system as a case study. Hence, this section will examine a pecan–cotton alley cropping system in Northwestern Florida, USA.

### *6.1 The system*

A 47-year-old pecan orchard was converted into an alley cropping system in 2001 (Fig. 3). The pecan trees were planted at a uniform spacing of 18.3 m and the orchard remained under grass cover for 29 years until it was converted into an alley cropping practice. The rows of trees were oriented in a north–south direction. Alleys measured 55 m in length and 18.3 m in width, with a practical cultivable width of 16.2 m. Before planting the cotton as the intercropped species, each alley was divided into two plots to assess the impacts of competition for water and nutrients on cotton production: barrier and non-barrier plots. The barrier plots were subjected to a root pruning treatment, in which a trenching machine was used to dig a 0.2 m wide and 1.2 m deep trench along both sides of the plot at a distance of 1.5 m from the trees to separate root systems of pecan and cotton. A double layer of 0.15-mm thick polyethylene sheeting was used to line the ditch prior to mechanical backfilling. The barrier plots thus served as the tree root exclusion treatment, preventing

**Figure 3** Pecan-cotton alley cropping system in Northwestern Florida, USA.

interaction of tree and cotton roots, while the non-barrier plots, which did not receive this treatment, served as the tree-crop competition treatment. Sixteen rows of cotton were planted in each alley. Measurements were made on rows 1, 4, 8, 9 and 16. The barrier and non-barrier treatment received standard fertilizer and pesticide application for cotton, and no irrigation was applied.

## 6.2 Production (yield)

Lint yield of cotton in the barrier treatment (677 kg ha$^{-1}$) was higher than that of the non-barrier treatment (502 kg ha$^{-1}$) but was not different from the monoculture stand (634 kg ha$^{-1}$). Within the non-barrier treatment, the highest yield was observed in the centre of the alley (row 8), with row 1 (at the tree-crop interface) yielding less than the centre row ($P = 0.0388$ for row 8; $P = 0.0216$ for row 9). The presence of the barrier had the greatest (positive) impact on plants in the two edge rows (1 and 16) of the barrier treatment (Allen et al. 2004a,b; Wanvestraut et al. 2004; Zamora et al. 2006, 2007, 2008). When high yield of the edge rows is removed (as they theoretically had no crop competitor on one side), the barrier treatment yield (556 kg ha$^{-1}$) was not different from the monoculture cotton yield (634 kg ha$^{-1}$). The vegetative development of cotton was strongly enhanced by the presence of the barrier treatment, which outperformed the non-barrier plants. Leaf area index in the barrier treatment was higher than that of the non-barrier plants. Although the absorbed PAR was 42% lower for the barrier plants compared to the monoculture plants, lint yield was similar for both treatments. This indicates that light was not a major factor influencing production in that system, but below ground competition for water and nutrients was present as evidenced by the significant reduction of yield in non-barrier treatment. Cotton tolerated moderate shading and provided acceptable yield when below ground competition was alleviated (Zamora et al. 2008).

## 6.3 Environmental impacts

Alley cropping is viewed as a possible strategy to clean up agricultural contaminants in groundwater. In conventional agricultural systems, less than half of the applied N and phosphorus fertilizer is taken up by the crops. Consequently, excess fertilizer is washed away from agricultural fields via surface run-off or leached into sub-surface water supply, thereby contaminating water sources and decreasing water quality (Cassman 1999; Jose 2009). Trees or shrubs with deep root systems can theoretically improve water quality by serving as a 'safety net' whereby excess nutrients are captured by tree roots and then recycled back into the system through root turnover and litter fall, increasing the nutrient use efficiency of the system (Allen et al. 2004a).

In the pecan-cotton alley cropping system, $NO_3$-N concentrations varied between the barrier and non-barrier treatments. At 0.3 and 0.9 m depths, $NO_3$-N levels in barrier treatments were recorded to be 14.36 and 10.82 mg $L^{-1}$, respectively, representing a 24.7% decrease with depth. Mean concentrations for non-barrier at corresponding depths were 10.16 and 7.15 mg $L^{-1}$, respectively, representing a 29.6% decrease. Furthermore, average $NO_3$-N leaching was higher in the barrier treatment (3.40 kg $ha^{-1}$ $month^{-1}$) than in the non-barrier treatment (1.28 kg $ha^{-1}$ $month^{-1}$). During the whole growing season, the amount of $NO_3$-N at 0.3 m depth in the barrier treatment was 121.94 kg $ha^{-1}$, which was significantly higher than in the non-barrier treatment (63.83 kg $ha^{-1}$), a decrease of 47.7% (Allen et al. 2004a,b). A similar trend occurred at 0.9 m depth, where cumulative $NO_3$-N leaching in barrier (45.56 kg $ha^{-1}$) was higher than in non-barrier (13.05 kg $ha^{-1}$), a 71.4% decrease. Therefore, these numbers indicate that the competitive presence of trees can be utilized to reduce N leaching in alley cropping systems.

Allen et al. (2004a) further observed competition for N in the pecan-cotton alley cropping system, where cotton plants in the barrier treatment had a 59% higher aboveground biomass compared to plants in the non-barrier treatment. Although a companion study indicated that competition for water was also a factor (Wanvestraut et al. 2004), the authors hypothesized that because pecan trees leaf out earlier in the spring and have a high nutrient demand early in the growing season, the soil N depleted before cotton plants were established later in the growing season. Therefore, in this particular system, cotton plants were more likely to rely on supplemental N to fulfil plant needs. In addition, Allen et al. (2005) showed that N mineralization rates differed between barrier and non-barrier treatments in this system. Higher rates of N mineralization were observed in the non-barrier treatment (26.05 mg $kg^{-1}$ $month^{-1}$) compared to rates observed in the barrier treatment (19.78 mg $kg^{-1}$ $month^{-1}$), indicating that competitive interactions for water and N in the non-barrier treatment may have led to decreased ability of cotton plants to uptake N (Allen et al. 2005; Jose et al. 2009).

Furthermore, integrating trees in the agricultural landscape using alley cropping can enhance soil physical, chemical and biological properties by adding significant amounts of above and below ground organic matter and thus releasing and recycling nutrients within the systems. Trees can increase nutrient inputs by retrieving nitrogen from deep soil horizons. Lee and Jose (2003a,b) reported a higher soil organic matter and microbial biomass level in the previously described pecan-cotton system compared to monocropped cotton. Generally, nutrients are lost in monoculture row crop systems due to total or substantial harvest of biomass that prevents the build-up of organic matter in the soil. Agroforestry's role in enhancing and maintaining long-term soil productivity and sustainability has been well documented (Nair 1993;

Young 1997; Schroth and Sinclair 2003). The incorporation of trees or shrubs in agroforestry systems can increase the amount of carbon sequestered compared to a monoculture field of crops (Sharrow and Ismail 2004; Kirby and Potvin 2007; Jose 2009). Carbon sequestered in agroforestry systems could be sold in carbon credit markets where such opportunities exist; so this is an example of how they may be capable of providing other long-term economic returns to farmers.

## 6.4 Adoptability

Alley cropping can contribute to achieving various environmental impacts. However, the success of agroforestry adoption in the landscape depends on how the practice can address many of the socioeconomic conditions of the farmers including enhancement of revenue, labour and machinery needs, time and effort to do it, among others. Although its potential as a sustainable agricultural practice exists, farmers are hesitant to adopt it due to lack of skills and scientific knowledge to implement it such as in the case for pecan-based alley cropping. To increase adoption of this practice, the system must be properly designed to address the economic bottom line of the farmers, and the adoption must conform to the farmer's available resources (e.g. machinery, manpower etc.). Furthermore, there is a need to understand factors influencing adoption behaviour, identifying impacts of land- and tree-tenure, developing methods for integrating socioeconomics and biophysical approaches to increase adoption. Understanding or improving agroforestry cost-benefit and impact analyses including incorporating of risk and uncertainty, estimating and incorporating non-market goods and services produced by agroforestry systems, multi-input and output production function analysis are keys to adoption of alley cropping systems.

Zinkhan and Mercer (1997), in their survey of land-use professionals in the Southern US and related literature review, found that silvopastoral systems were the most common type of agroforestry in the region, which is also similar in Europe (Mosquera-Losada et al. 2016). In this context, the most frequently mentioned benefits associated with these systems included the potential for increased economic returns, diversification and the enhancement of the timing of cash flows. However, more study is needed on the verification of economic benefits from various temperate agroforestry systems, including ground-level agricultural economics studies. These are by nature very situation-specific and include various economic inputs and outputs that must be weighed against monocrop production benefits and farmer economic priorities.

Lastly, it should be noted that, in today's highly competitive and market-driven economy, caution must be observed when considering large-scale adoption of diverse cropping systems, as economic path dependency associated with intensive agricultural production (such as in the US Corn Belt) can work against the incorporation of diverse crops, as described by Roesch-McNally et al. (2018). They conclude that this issue needs to be further addressed by adjusting policy and economic incentives that foster diverse systems and reduce climate risk.

## 7 Future trends and conclusion

Considered as a holy grail in agriculture, properly designed alley cropping has shown promise to achieve the cornerstones of sustainable agriculture in temperate regions compared to conventional agriculture. Alley cropping can promote economic productivity while enhancing ecological integrity. As a perennial system, enhancing the positive interactions between the tree or shrub and crop while minimizing the negative ones can enhance economic production. Even though crop yield may be decreased when compared to monoculture systems, by incorporating multiple species in a single system there is a possibility to increase overall biomass production and economic value. Environmentally, the trees in alley cropping could play the safety-net role, which shows the potential improvement of water quality by reducing surface and sub-surface contamination.

Although the adoption of the practice is still slow, the growing intensity of land-use practices, the increasing public interest in achieving ecosystem services (clean water, wildlife habitat, carbon sequestration, among others) all serve to encourage farmers to evaluate the potential of alley cropping in their farm operation to meet their land management needs. Alley cropping can advance temperate zone stewardship by (1) converting degraded or marginal land into being productive estate, (2) protecting sensitive lands and (3) enhancing, diversifying and promoting farm production systems.

Temperate alley cropping research is still limited, and several information gaps still exist that need immediate attention by the scientific community. There is a need to quantify the interactive effects of trees and crops at landscape level and at different geographic scales. Furthermore, appropriate design of alley cropping systems at various levels and scales needs to be explored. Likewise, new technologies stemming from GPS data, precision agriculture, online weather databases and Landsat-type satellite imagery need to be incorporated into land-use decision making. Also, appropriate incentives and policies need to be pursued in order to create an environment where alternative cropping systems are financially viable and geared towards long-term climate change response. In sum, while there are many challenges to the adoption of alley cropping and other alternative systems, there are also many opportunities, as more and more landowners and consumers recognize the importance of achieving mutual sustainability in the agricultural and environmental landscapes.

## 8 Where to look for further information

### *National Agroforestry Center: https://www.fs.usda.gov/nac/practices/alleycropping.shtml*

The USDA National Agroforestry Center (NAC) accelerates the application of agroforestry through a national network of partners. NAC conducts research, develops technologies and tools, coordinates demonstrations and training, and provides useful information to natural resource professionals on agroforestry.

### *The Center for Agroforestry at the University of Missouri: http://www.centerforagroforestry.org, Alley Cropping: http://www.centerforagroforestry.org/practices/ac.php*

The Center for Agroforestry at the University of Missouri, established in 1998, is one the world's leading centres contributing to the science underlying agroforestry. The faculty at the Center has a number of members from various disciplines doing research and outreach activities on agroforestry around the United States.

### *AGFORWARD: http://www.agforward.eu/index.php/en/silvoarable-agroforestry-in-the-uk.html*

AGFORWARD (AGroFORestry that Will Advance Rural Development) is a four-year research project funded by the EU's Seventh Framework Program for Research and Technological Development. The project builds on existing agroforestry experiments, current on-farm agroforestry trials and previous research projects such as 'Silvoarable Agroforestry For Europe (SAFE)'. The European Agroforestry Federation is a partner. The start of the project coincides with the launch of EU Rural Development Regulations that can support the establishment of agroforestry systems.

### *Association for Temperate Agroforestry (AFTA): http://www.aftaweb.org/*

AFTA is a group of natural resource professionals, researchers and farmers whose mission is to promote the wider adoption of agroforestry by landowners in temperate regions of North America. With members in the United States, Canada and overseas, AFTA pursues its mission through activities such as networking, information exchange, public education and policy development.

### *Savanna Institute: http://www.savannainstitute.org/*

The Savanna Institute is a nonprofit organization working to lay the groundwork for widespread adoption of agroforestry in the Midwest US. The Institute works in collaboration with farmers and scientists to develop perennial food and fodder crops within multifunctional polyculture systems grounded in ecology and inspired by the savanna biome. It strategically enacts this mission via research, education and outreach.

## 9 References

Allen, S., Jose, S., Nair, P. K. R. and Brecke, B. J. 2004a. Competition for $^{15}N$ labeled nitrogen in a pecan-cotton alley cropping system in the southern United States. *Plant and Soil* 263:151-64.

Allen, S., Jose, S., Nair, P. K. R., Brecke, B. J. and Nkedi-Kizza, P. 2004b. Safety net role of tree roots: Experimental evidence from an alley cropping system. *Forest Ecology and Management* 192:395-407.

Allen, S., Jose, S., Nair, P. K. R., Brecke, B. J., Nair, V. D., Graetz, D. and Ramsey, C. L. 2005. Nitrogen mineralization in a pecan (*Carya illinoensis* K. Koch)-cotton (*Gossypium hirsutum* L.) alley cropping system in the southern United States. *Biology and Fertility of Soils* 41:28-37.

Altieri, M. A. 1983. *Agroecology: The Scientific Basis of Alternative Agriculture*. Golden Horn Press, Berkeley, CA.

Bakermans, M., Rodewald, A., Vitz, A. and Rengifo, C. 2012. Migratory bird use of shade coffee: The role of structural and floristic features. *Agroforestry Systems* 85:85-94.

Best, L. B., Whitmore, R. C. and Booth, G. M. 1990. Use of cornfields by birds during the breeding season. The importance of edge habitat. *American Midland Naturalist* 123:84-99.

Bonilla, C. A., Munoz, J. F. and Vauclin, M. 1999. Opus simulation of water dynamics and nitrate transport in a field plot. *Ecological Modeling* 122:69-80.

Browaldh, M. 1995. The influence of trees on N dynamics in an agrisilvicultural system in Sweden. *Agroforestry Systems* 30:131-8.

Bugg, R. L., Sarrantonio, M., Dutcher, J. D. and Phatak, S. C. 1991. Understory cover crops in pecan orchards: Possible management systems. *American Journal of Alternative Agriculture* 6:50-62.

Burgess, S. O., Adams, M. A., Turner, N. C. and Ong, C. K. 1998. The redistribution of soil water by tree roots systems. *Oecologia* 115:306-11.

Caldwell, M. M. and Richards, J. H. 1989. Hydraulic lift: Water efflux from upper roots improves effectiveness of water uptake by deep roots. *Oecologia* 79:1-5.

Cassman, K. G.1999. Ecological intensification of cereal production systems: Yield potential, soil quality, and precision agriculture. *Proceedings of the National Academy of Sciences* 96:5952-9.

Corak, S. J., Blevins, D. G. and Pallardy, S. G. 1987. Water transfer in an alfalfa-maize association: Survival of maize during drought. *Plant Physiology* 84:582-6.

Dawson, I. K., Place, F., Torquebiau, E., Malézieux, E., Iiyama, M., Sileshi, G. W., Kehlenbeck, K., Masters, E., McMullin, E. and Jamnadass, R. 2013. Agroforestry, food and nutritional security. *FAO*. http://www.fao.org/forestry/37082-04957fe26afbc90d1e9c0356c48185295.pdf (Last accessed 22 January 2018).

Dyack, B., Rollins, K. and Gordon, A. M. 1999. An economic analysis of market and non-market benefits of a temperate intercropping system in southern Ontario, Canada. *Agroforestry Systems* 44:197-214.

Eckberg, J., Johnson, G., Heimpel, G., Sheaffer, C., Peterson, J., Plecas, M., Kaser, J. and Wyse, D. 2015. Integrative cropping systems to enhance biological control and increase soybean yield. *University of Minnesota - Monsanto Fellows Meeting*, 29–30 January.

Eichhorn, M. P., Paris, P., Herzog, F., Incoll, L. D., Liagre, F., Mantzanas, K., Mayus, M., Moreno, G., Papanastasis, V. P., Pilbeam, D. J., Pisanelli, A. and Dupraz, C. 2006. Silvoarable systems in Europe - past, present and future prospects. *Agroforestry Systems* 67:29–50.

FAO. 2013. Advancing agroforestry on the policy agenda: A guide for decision-makers. Agroforestry Working Paper No.1. FAO, Rome.

Garrett, H. E. and Buck, L. E. 1997. Agroforestry practice and policy in the United States of America. *Forest Ecology and Management* 91:5–15.

Garrett, H. E. and Harper, L. S. 1999. The science and practice of black walnut agroforestry in Missouri, USA: A temperate zone assessment. *In* Buck, L. E., Lassoie, J. P. and Fernandes, E. C. M. (Eds), *Agroforestry in Sustainable Agriculture Systems*. CRC Press, New York, NY, pp. 97–110.

Garrett, H. E., McGraw, R. L. and Walter, W. D. 2009. Alley cropping practices. *In* Garrett, H. E. (Ed.), *North American Agroforestry: An Integrated Science and Practice*, 2nd Edition. American Society of Agronomy, Madison, WI.

Garrity, D. 2004. Agroforestry and the achievement of the millennium development goals. *Agroforestry Systems* 61:5–17.

Gibbs, S., Koblents, H., Coleman, B., Gordon, A., Thevathasan, N. and Williams, P. 2016. Avian diversity in a temperate tree-based intercropping from inception to now. *Agroforestry Systems* 90:905–16.

Gillespie, A. R., Miller, B. K. and Johnson, K. D. 1995. Effects of ground cover on tree survival and growth in filter strips of the Cornbelt Region of the midwestern US. *Agriculture, Ecosystems and Environment* 53:263-70.

Gillespie, A. R., Jose, S., Mengel, D. B., Hoover, W. L., Pope, P. E., Seifert, J. R., Biehle, D. J., Stall, T. and Benjamin, T. 2000. Defining competition vectors in a temperate alleycropping system in the midwestern USA: I. Production physiology. *Agroforestry Systems* 48:25–40.

Girma, H., Rao, M. R. and Sithanantham, S. 2000. Insect pests and beneficial arthropod populations under different hedgerow intercropping in semiarid Kenya. *Agroforestry Systems* 50:279–92.

Gordon, A. M. and Williams, P. A. 1991. Intercropping valuable hardwood tree species and agricultural crops in southern Ontario. *The Forestry Chronicle* 67:200–8.

Gordon, A. M., Newman, S. M. and Williams, P. A. 1997. *Temperate Agroforestry System: An Overview*. CAB International, Wallingford, UK, pp. 1–6.

Govindarajan, M., Rao, M. R., Mathuva, M. N. and Nair, P. K. R. 1996. Soil-water and root dynamics under hedgerow intercropping in semiarid Kenya. *Agronomy Journal* 88:513-20.

Hagan, D., Jose, S., Thetford, M. and Bohn, K. 2009. Production physiology of three native shrubs interplanted in a young longleaf pine plantation. *Agroforestry Systems* 76:283–94

Holzmueller, E. J. and Jose, S. 2012. Bioenergy crops in agroforestry systems: Potential for the U.S. North Central Region. *Agroforestry Systems* 85:305–14

Horton, J. L. and Hart, S. C. 1998. Hydraulic lift: A potentially important ecosystem process. *Trends in Ecological Evolution* 13:232-5.

Inderjit, and Malik, A. U. 2002. *Chemical Ecology of Plants: Allelopathy in Aquatic and Terrestrial Ecosystems*. Birkhauser-Verlag, Berlin, Germany, 272pp.

Jose, S. 2009. Agroforestry for ecosystem services and environmental benefits: An overview. *Agroforestry Systems* 76:1–10.

Jose, S. 2011. Managing native and non-native plant species in agroforestry. *Agroforestry Systems* 83:101–5.

Jose, S. and Gillespie, A. R. 1998a. Allelopathy in black walnut (*Juglans nigra* L.) alley cropping: I. Spatio-temporal variation in soil juglone in a black walnut-corn (*Zea mays* L.) alley cropping system. *Plant and Soil* 203:191–7.

Jose, S. and Gillespie, A. R. 1998b. Allelopathy in black walnut (*Juglans nigra* L.) alley cropping. II. Effects of juglone on hydroponically grown corn (*Zea mays* L.) and soybean (*Glycine max* (L.) Merr.) growth and physiology. *Plant and Soil* 203:199–205.

Jose, S. and Holzmueller, E. 2008. Black walnut allelopathy: Implications for intercropping. *In* Zeng, R. S., Mallik, A. U. and Luo, S. M. (Eds), *Allelopathy in Sustainable Agriculture and Forestry*. Springer, New York, NY.

Jose, S., Gillespie, A. R., Seifert, J. R. and Biehle, D. J. 2000a. Defining competition vectors in a temperate alleycropping system in the midwestern USA: II. Competition for water. *Agroforestry Systems* 48:41–59.

Jose, S., Gillespie, A. R., Seifert, J. E. and Pope, P. E. 2000b. Defining competition vectors in a temperate alleycropping system in the midwestern USA: III. Competition for nitrogen and litter decomposition dynamics. *Agroforestry Systems* 48:61–77.

Jose, S., Gillespie, A. R. and Pallardy, S. G. 2004. Interspecific interactions in temperate agroforestry. *Agroforestry Systems* 61:237–55.

Jose, S., Holzmueller, E. J. and Gillespie, A. R. 2009. Tree-crop interactions in temperate agroforestry. *In* Garrett, H. E. (Ed.), *North American Agroforestry: An Integrated Science and Practice*, 2nd Edition. American Society of Agronomy, Madison, WI. Chapter 4.

Jose, S., Gold, M. A. and Garrett, H. E. 2012. The future of temperate agroforestry in the United States. *In* Garrity, D. and Nair, P. K. R. (Eds), *Agroforestry: The Way Forward*. Springer Science, the Netherlands.

Kirby, K. R. and Potvin, C. 2007. Variation in carbon storage among tree species: Implications for the management of a small-scale carbon sink project. *Forest Ecology and Management* 246:208–21.

Lee, K. H. and Jose, S. 2003a. Soil respiration, fine root production and microbial biomass in cottonwood and loblolly pine plantations along a soil nitrogen gradient. *Forest Ecology and Management* 185:263–73.

Lee, K. H. and Jose, S. 2003b. Soil respiration and microbial biomass in a pecan-cotton alley cropping system in southern USA. *Agroforestry Systems* 58:45–54.

Lehmann, J., Peter, I., Steglich, C., Gebauer, G., Huwe, B. and Zech, W. 1998. Belowground interactions in dryland agroforestry. *Forest Ecology and Management* 111:157–69.

Lewis, C. E., Tanner, G. W. and Terry, W. S. 1985. Double vs. single-row pine plantations for wood and forage production. *Southern Journal of Applied Forestry* 9:55–61.

Lin, C. H., McGraw, R. L., George, M. F. and Garrett, H. E. 1999. Shade effects on forage crops with potential in temperate agroforestry practices. *Agroforestry Systems* 44:109–19.

MacLeod, C., Blackwell, G. and Benge, J. 2012. Reduced pesticide toxicity and increased woody vegetation cover account for enhanced native bird densities in organic orchards: Orchard management and native birds. *Journal of Applied Ecology* 49:652–60.

Magcale-Macandog, D. B. M., Ranola, F. M., Ranola, R. and Vidal, P. A. B. 2010. Enhancing the food security of upland farming households through agroforestry in Claveria, Misamis Oriental Philippines. *Agroforestry Systems* 79:327–42.

Malézieux, E., Crozat, Y., Dupraz, C., Laurans, M., Makowski, D., Ozier-Lafontaine, H., Rapidel, B., de Tourdonnet, S. and Valantin-Morison, M. 2009. Mixing plant species in cropping systems: Concepts, tools and models. A review. *Agronomy for Sustainable Development* 29:43–62.

Marshall, M. G. and Bennett, C. F. (Eds). 1998. *Outcomes of Nitrogen Fertilizer Management Programs, Vol. 3. National Extension Targeted Water Quality Program, 1992–1995*. Texas A&M University, State College, TX, Cooperating with Cooperative State Research, Education, and Extension Service (CSREES), USDA.

Miller, A. W. and Pallardy, S. G. 2001. Resource competition across the crop-tree interface in a maize-silver maple temperate alley cropping stand in Missouri. *Agroforestry Systems* 53:247–59.

Molnar, T., Kahn, P., Ford, T., Funk, C. and Funk, C. 2013. Tree crops, a permanent agriculture: concepts from the past for a sustainable future. *Resources* 2, 457–88.

Monteith, J. L., Ong, C. K. and Corlett, J. E. 1991. Microclimatic interactions in agroforestry. *Forest Ecology and Management* 45:31–44.

Mosquera-Losada, M. R., Santiago Freijanes, J. J., Pisanelli, A., Rois, M., Smith, J., den Herder, M., Moreno, G., Malignier, N., Mirazo, J. R., Lamersdorf, N., Ferreiro Domínguez, N., Balaguer, F., Pantera, A., Rigueiro-Rodríguez, A., Gonzalez-Hernández, P., Fernández-Lorenzo, J. L., Romero-Franco, R., Chalmin, A., Garcia de Jalon, S., Garnett, K., Graves, A. and Burgess, P. J. 2016. Extent and success of current policy measures to promote agroforestry across Europe. Deliverable 8.23 for EU FP7 Research Project: AGFORWARD 613520, 95pp.

Nair, P. K. R. 1993. *An Introduction to Agroforestry*. Kluwer Academic Publishers, Dordrecht, the Netherlands, pp. 499.

NeSmith, D. S. and Ritchie, J. T. 1992. Short- and long-term responses of corn to pre-anthesis soil water deficit. *Agronomy Journal* 84:107–13.

Ng, H. Y. F., Drury, C. F., Serem, V. K., Tan, C. S. and Gaynor, J. D. 2000. Modeling and testing of the effect of tillage, cropping and water management practices on nitrate leaching in clay loam soil. *Agricultural Water Management* 43:111–31.

Ong, C. K., Deans, J. D., Wilson, J., Mutua, J., Khan, A. A. H. and Lawson, E. M. 1999. Exploring belowground complementarity in agroforestry using sap flow and root fractal techniques. *Agroforestry Systems* 44:87–103.

Palma, J., Graves, A. R., Burgess, P. J., van der Werf, W. and Herzog, F. 2007. Integrating environmental and economic performance to assess modern silvoarable agroforestry in Europe. *Ecological Economics* 63:759–67.

Pardini, A., Longhi, F. and Natali, F. 2008. Pastoral systems and agro-tourism in marginal areas of Central Italy. *Options Méditerranéennes* 79:97–102.

Pekin, B. and Pijanowski, B. 2012. Global land use intensity and the endangerment status of mammal species. *Diversity and Distribution* 18:909–18.

Peng, K., Incoll, R. D., Sutton, L. L., Wright, S. and Chadwick, C. 1993. Diversity of airborne arthropods in a silvoarable agroforestry system. *Journal of Applied Ecology* 30 551–60.

Pimentel, D. Harvey, C. Resosudarmo, P., Sinclair, K., Kurz, D., McNair, M., Crist, S., Shpritz, L., Fitton, R., Saffouri, R. and Blair, R. 1995. Environmental and economic costs of soil erosion and conservation benefits. *Science* 267:1117–23.

Regueiro, R. A., McAdam, J. and Mosquera-Lozada, M. R. 2009. *Agroforestry in Europe: Current Status and Future Prospects*. Springer Science Business Media B.V., Dordrecht, the Netherlands, pp. 295–320.

Renken, R. B. 1983. Breeding bird communities and bird habitat associations on North Dakota waterfowl production areas of three habitat types. Unpublished MS Thesis, Iowa State University, Ames, IA.

Reynolds, P. E., Simpson, J. A., Thevathasan, N and Gordon, A. M. 2007. Effects of tree competition on corn and soybean photosynthesis, growth, and yield in a temperate tree-based agroforestry intercropping system in southern Ontario, Canada. *Ecological Engineering* 29:362–71.

Roesch-McNally, G. E., Arbuckle, J. G. and Tyndall, J. C. 2018. Barriers to implementing climate resilient agricultural strategies: The case of crop diversification in the U.S. Corn Belt. *Global Environmental Change* 48:206–15.

Root, R. 1973. Organization of a plant-arthropod association in simple and diverse habitats: The fauna of collards (*Brassica oleracea*). *Ecological Monographs* 43:95–124.

Rowe, E. C., Hairiah, K, Giller, K. E., van Noordwijk, M. and Cadisch, G. 1999. Testing the safety-net role of hedgerow tree roots by $^{15}N$ placement at different soil depths. *Agroforestry Systems* 43:81–93.

Schroth, G and Sinclair, F. 2003. *Trees Crops and Soil Fertility: Concepts and Research Methods*. CABI, Wallingford, UK, p. 464.

Sharrow, S. H. and Ismail, S. 2004. Carbon and nitrogen storage in agroforests, tree plantations, and pastures in western Oregon, USA. *Agroforestry Systems* 60:123–30.

Smith, J., Pearce, B. D. and Wolfe, M. 2013. Reconciling productivity with protection of the environment: Is temperate agroforestry the answer? *Renewable Agriculture and Food Systems* 28(1):80–92.

Stamps, W. T. and Linit, M. J. 1998. Plant diversity and arthropod communities: Implications for temperate agroforestry. *Agroforestry Systems* 39:73–89.

Stamps, W. T., Wood, T. W., Linit, M. G. J. and Garrett, H. E. 2002. Arthropod diversity in alley cropped black walnut (*Juglans nigra* L.) stands in eastern Missouri, USA. *Agroforestry Systems* 56:167–75.

Thevathasan, N. V. and Gordon, A. M. 2004. Ecology of tree intercropping systems in north temperate region: Experience from southern Ontario, Canada. *Agroforestry Systems* 61:257–68.

Thevathasan, N. V., Gordon, A. M. and Voroney, R. P. 1998. Juglone (5-hydroxy-1,4 napthoquinone) and soil nitrogen transformation interactions under a walnut plantation in southern Ontario, Canada. *Agroforestry Systems* 44:51–162.

Udawatta, R. P., Nygren, P. and Garett, H. E. 2005. Growth of three oak species during establishment of an agroforestry practice for watershed protection. *Canadian Journal of Forestry Research* 35:602–9.

University of Missouri Center for Agroforestry (UMCA). 2015. Alley cropping. *In Training Manual for Applied Agroforestry Practices*. Columbia, MO. http://www.centerforagroforestry.org/pubs/training/index.php

USDA. 2018. Agricultural census. https://www.agcensus.usda.gov/ (Last accessed 22 January 2018).

USDA NRCS. 2007. Soil erosion on cropland. https://www.nrcs.usda.gov/wps/portal/nrcs/detail/national/home/?cid=stelprdb1041887 (Last accessed 22 January 2018).

Vandermeer, J. 1989. *The Ecology of Intercropping*. Cambridge University Press, Cambridge, UK, 249pp.

van Noordwijk, M., Lawson, G., Soumaré, A., Groot, J. J. R. and Hairiah, K. 1996. Root distribution of trees and crops: Competition and/or complementarity. *In* Ong,

C. K. and Huxley, P. (Eds), *Tree-Crop Interactions: A Physiological Approach*. CAB International, Wallingford, UK, pp. 319-64.

Vasileiadis, V., Moonen, A., Sattin, M., Otto, S., Pons, X., Kudsl, P., Veres, A., Dorner, Z., van der Weide, R., Marraccini, E., Pelzer, E., Angevin, F. and Kiss, J. 2013. Sustainability of European maize-based cropping systems: Economic, environmental and social assessment of current and proposed innovative IPM-based systems. *European Journal of Agronomy* 48:1-11.

Wanvestraut, R., Jose, S., Nair, P. K. R. and Brecke, B. J. 2004. Competition for water in a pecan-cotton alley cropping system in the southern United States. *Agroforestry Systems* 60:167-79.

Wilcove, D., Rothstein, D., Dubow, J., Phillips, A. and Losos, E. 1998. Quantifying threats to imperiled species in the United States. *Bioscience* 48:607-15.

Wilson, M. 2007. Perennial pathways: Planning and establishment practices for edible agroforestry. Master of Science Thesis, University of Illinois, Urbana, IL.

Workman, S. W., Bannister, M. E. and Nair, P. K. R. 2003. Agroforestry potential in the southeastern United States: Perceptions of landowners and extension professionals. *Agroforestry Systems* 59:73-83.

Yahner, R. 1982. Avian use of vertical strata and plantings in farmstead shelterbelts. *Journal of Wildlife Management* 46:50-60.

Young, A. 1997. *Agroforestry for Soil Management*, 2nd ed. CAB International, Wallingford, UK, p. 307.

Yunusa, I. A. M., Thomson, S. E., Pollock, K. P., Youwei, L and Mead, D. J. 2005. Water potential and gas exchange did not reflect performance of *Pinus radiata* D. Don in an agroforestry system under conditions of soil-water deficit in a temperate environment. *Plant and Soil* 275:195-206.

Zamora, D., Jose, S. and Nair, P. K. R. 2006. Interspecific interaction in a pecan-cotton alleycropping system in the southern United States: The production physiology. *Canadian Journal of Botany* 84:1686-94.

Zamora, D., Jose, S. and Nair, P. K. R. 2007. Morphological plasticity of cotton roots in response to interspecific competition with pecan in an alleycropping system in the southern United States. *Agroforestry Systems* 69:107-16.

Zamora, D. S., Jose, S., Nair, P. K. R., Jones, J. W., Brecke, B. J. and Ramsey, C. L. 2008. Interspecific competition in a pecan-cotton alleycropping system in the southern United States: Is light the limiting factor?. *In* Jose, S. and Gordon, A. M. (Eds), *Toward Agroforestry Design: An Ecological Approach*. Springer Science, the Netherlands, pp. 81-95.

Zamora, D., Jose, S. and Napolitano, K. 2009a. Competition for $^{15}N$ labeled nitrogen in a loblolly pine-cotton alleycropping system in the southern United States. *Agriculture, Ecosystems and Environment* 131:40-50.

Zamora, D., Jose, S., Jones, J. W. and Cropper, W. P. 2009b. Modeling cotton production in a pecan alleycropping system using CROPGRO. *Agroforestry Systems* 76:423-35.

Zinkhan, F. C. and Mercer, D. E. 1997. An assessment of agroforestry systems in southern USA. *Agroforestry Systems* 35:303-21.

# Chapter 4

# Sustainable production of willow for biofuel use

*M. Weih, P.-A. Hansson, J. A. Ohlsson, M. Sandgren, A. Schnürer and A.-C. Rönnberg-Wästljung, Swedish University of Agricultural Sciences, Sweden*



## 1 Introduction

Willows (genus *Salix*, family Salicaceae) are fast-growing deciduous trees and shrubs occurring mostly in temperate and arctic zones of the northern hemisphere (Dickmann and Kuzovkina, 2014). Traditionally, various species and cultivars of willows have been used in many countries of the world for centuries, for example, for basket-making, fencing, and small-scale husbandry. More recently, intensively managed plantations of willows and other fast-growing trees on agricultural land are gaining increasing interest in many countries of the world, mainly due to their efficient and sustainable land use in combination with an increasing demand for biofuel sources (Karp and Shield, 2008). The largest willow plantations for industrial, biofuel, and environmental purposes are currently found in China, Argentina, North America, and some European countries (FAO, 2016) (Fig. 1). In some of these countries, the shoots from

http://dx.doi.org/10.19103/AS.2019.0027.17

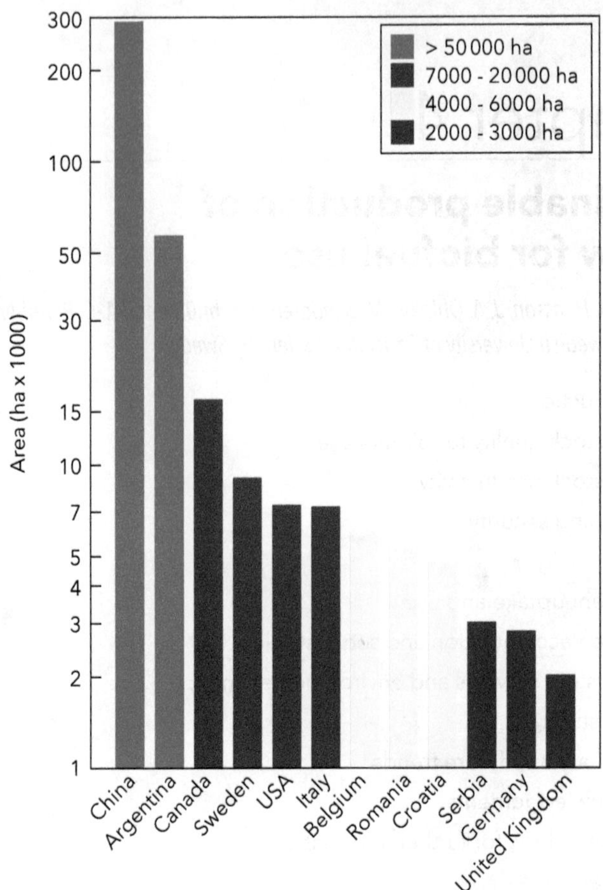


**Figure 1** Areas of planted willow (*Salix* spp.) worldwide in 2015 (FAO, 2016). Only countries with willow plantation areas ≥2000 ha are shown. Note the $\log_e$ scale of the *y*-axis.

willows have been used commercially as biofuel in the form of wood chips for direct combustion in heat and power plants since the late 1980s (Kuzovkina et al., 2008) (Fig. 2). There is now also an increasing interest of using biomass from willows as raw material for conversion into biofuels such as biogas and bioethanol (Karp et al., 2011; Phitsuwan et al., 2013), a development that requires an enhanced focus in research and development on feedstock quality when willow is to be used as raw material for different supply chains.

Willow production for biofuel is mostly based on an intensive production system known as short rotation coppice (SRC) system (Larsson et al., 2007). The SRC cultivation practice is based on intensive site preparation prior to planting, including weed control, mechanical planting, in many cases mineral nutrient

**Figure 2** Commercial plantations of willow for biofuel are a common view in the agricultural landscape of South and Central Sweden. Photos: M. Weih (left) and J. Svennås-Gillner (right), SLU, Sweden.

fertilizer inputs at the beginning of each cutting cycle, and multiple harvests on the basis of 2- to 5-year cutting cycles (Fig. 3) (Weih, 2013; Verwijst et al., 2013). The plantations consist of densely grown (from cuttings), high-yielding varieties of willow cultivated on agricultural land at densities of 10 000–20 000 plants ha$^{-1}$. The shoots are typically harvested during winter and processed to wood chips on-site by using specialized machinery (Fig. 4). The rootstock or stools remain in the ground after harvest with new shoots resprouting the following spring. A plantation could be viable for more than 20 years before reestablishment becomes necessary.

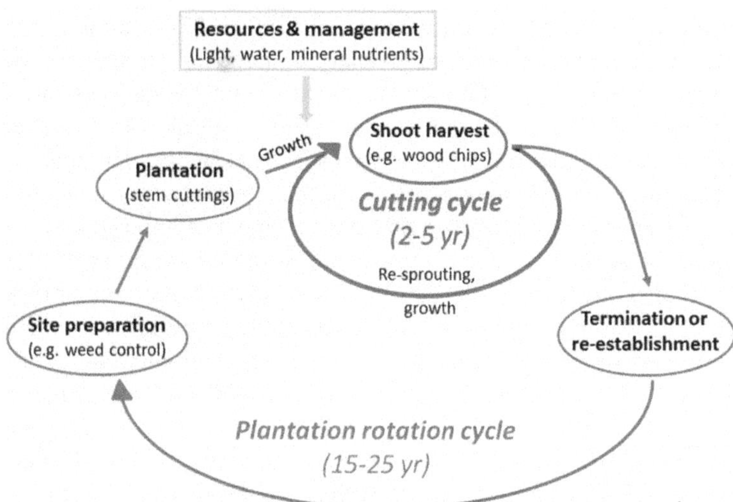


**Figure 3** Overview of the actions and processes in a typical willow (*Salix*) short rotation coppice plantation grown for biofuel. Source: redrawn and modified after Weih (2013).

**Figure 4** Willow plantations for direct combustion of biomass in heat and power plants are harvested during winter with specialized machinery that both cuts the rods and processes them directly into wood chips, which are supplied to the end user. Photo: N.-E. Nordh, SLU, Sweden.

The cultivation of plantations for biofuel use is often motivated in terms of beneficial effects on the environment and sustainable development (Karp and Shield, 2008). A major challenge is striking the balance between long-term ecological sustainability and reaching short-term crop productivity goals. In terms of sustainable development, the overall framework is here the document known as the Brundtland report (WCED, 1987), according to which sustainable development involves meeting the needs of the present without compromising the ability of future generations to meet their own needs. In this chapter we only focus on the primary dimension of sustainable development, that is, safeguarding long-term ecological sustainability (Holden et al., 2014). Ecological sustainability in biofuel production involves minimal chemical inputs, efficient nutrient recycling, and enhancement of important ecological processes related to nutrient acquisition, decomposition, and cropping security (e.g. protection against pests and pathogens), that is, those processes that often appear to be stimulated by increased biodiversity. In a resource-use perspective, ecological sustainability in the production and use of biofuels could aim at enhanced biofuel output with maintained or reduced depletion of natural resources (Weih et al., 2014).

We consider the following key issues for the sustainable production and use of willow for biofuel: the quality of the feedstock to be used as biofuel, feedstock productivity, cropping security, biodiversity, nutrient uptake and use, carbon accumulation and sequestration, and ecosystem services and environmental impact. Crop improvement programs in terms of selection and breeding mostly target feedstock productivity and quality traits at the individual plant scale, whereas several of the above key issues only can be evaluated at stand or landscape scales. This chapter provides an overview of the main challenges and key issues in the sustainable production and use of willow as raw material for biofuel production in temperate regions, and links crop improvement issues acting at individual plant scale to ecological sustainability issues frequently operating at higher scales.

## 2 Feedstock quality for biofuel use

In addition to the traditional use of willow biomass as biofuel for the production of heat and electricity by direct combustion (Ledin, 1996), there is recently a growing interest for the conversion of willow biomass into transportation fuels, with a strong focus on biogas and bioethanol (Sassner et al., 2008b; Serapiglia et al., 2013b; Pawar et al., 2018; Horn et al., 2011; Estevez et al., 2014). The biogas and bioethanol are produced by exposing the biomass to various microbiological processes. In bioethanol production, the cellulosic carbohydrates in the biomass are converted into fermentable monomers through an initial pretreatment, typically chemical or thermal, followed by enzymatic digestion. The monomers are further converted by yeast to ethanol, either simultaneously with enzymatic digestion or in separate steps (Sassner et al., 2008b). In biogas production, the biomass is degraded by a consortium of microbes under anaerobic conditions, known as anaerobic digestion. This process can, in theory, proceed without any pretreatment; however, a pretreatment breaking up the lignocellulosic structure improves accessibility and can increase both biogas yields and production rates (Horn et al., 2011; Estevez et al., 2012; Alexandropoulou et al., 2017).

For bioethanol production, the most important parameters are arguably the cellulose content of the biomass and secondary cell wall resistance (commonly referred to as biomass recalcitrance). For biogas, a high cellulose content and/or low lignin content of the willow biomass is desirable to achieve high biogas yields (Pawar et al., 2018). In general, quantities of major chemical components in willow shoot biomass vary greatly depending on the environment in which the feedstock is grown, feedstock genetics (species and variety), and also the method applied for their determination (Table 1). For example in one study, cellulose was found to be more variable than other biomass components in willow, but most of the variability was due to the environment and not the willow genotype (Fabio et al., 2017). Also biomass recalcitrance is highly variable in willow (Serapiglia et al., 2013b; Ray et al., 2012). Recently, several high-throughput screening facilities for rapid assessment of biomass recalcitrance have been constructed (Decker et al., 2009). These facilities enable the

**Table 1** Major chemical components (%) of willow shoot biomass reported in the literature. Cellulose and hemicellulose values may be reported as glucan and xylan, respectively, in the reference but are treated interchangeably here

| Cellulose | Hemicellulose | Lignin | Ash | Method | Reference |
|---|---|---|---|---|---|
| 55.9 (0.5) | 14.0 (1.0) | 13.8 (0.4) | 1.3 (0.1) | Wet chemistry | Szczukowski et al. (2002)[a] |
| 41.5 | 15.0 | 25.2 | 1.9 | Wet chemistry | Sassner et al. (2008a) |
| 39.7 (1.8) | 22.5 (1.6) | 22.7 (1.6) | N/A | Wet chemistry | Serapiglia et al. (2009) |
| 47.6 (3.1) | 28.3 (3.4) | 24.7 (1.4) | N/A | Wet chemistry | Sandak and Sandak (2011) |
| 42.5 (0.4) | 26.1 (0.5) | 23.0 (0.3) | N/A | Wet chemistry | Stolarski et al. (2011) |
| 34.8–41.8 | 12.2–15.2 | 23.9–28.8 | N/A | Wet chemistry | Ray et al. (2012) |
| 44.1 (0.2) | 21.8 (0.1) | 20.4 (0.3) | 2.1 (0.1) | Wet chemistry | Stolarski et al. (2013) |
| 41.6 (3.3) | 14.0 (1.2) | 28.0 (2.0) | 1.9 (0.3) | Wet chemistry | Serapiglia et al. (2015) |
| 44.4 (1.7) | 31.0 (1.2) | 25.5 (0.7) | 1.3 (0.2) | Wet chemistry | Krzyzaniak et al. (2014) |
| 39–41 | 32–33 | 22–23 | N/A | HR-TGA | Serapiglia et al. (2012)[b] |
| 38.7–44.5 | 30.1–33.6 | 20.9–24.1 | 0.5–4.5 | HR-TGA | Serapiglia et al. (2013a) |
| 38.4–45.3 | 31.1–34.9 | 20.3–23.2 | N/A | HR-TGA | Serapiglia et al. (2013b) |
| 41.3–44.0 | 18.3–19.4 | 25.0–25.9 | 1.6–2.7 | HR-TGA | Fabio et al. (2017)[c] |
| 37.7 (1.8) | 33.4 (2.3) | 25.6 (1.2) | N/A | NIR spectroscopy | Sandak et al. (2017) |

[a] Data refers to 3-year-old shoots.
[b] Values estimated from figures.
[c] Range refers to genotypic means.
Values are reported either as ranges or as means with standard deviations in parentheses, and in one case only a single value was reported. N/A means the value was not reported.
Abbreviations: HR-TGA, high-resolution thermogravimetric analysis; NIR, near infrared.

screening of large numbers of biomass samples for recalcitrance characteristics by breeders and researchers with an interest in elucidating the genetics and functional mechanisms underlying recalcitrance; a recent example is the screening of recalcitrance characteristics in poplars, which are taxonomically closely related to willows (Bhagia et al., 2016).

Independent of the biomass conversion process in consideration, the chemical composition of the biomass strongly influences the final biofuel yield. However, the most desirable chemical composition of the feedstock strongly depends on the supply chain targeted. Besides the biomass composition influencing the amount of output of the targeted product (e.g., transportation fuel or energy from direct combustion) per unit of biomass, the most important factor affecting biofuel yield is the productivity of the feedstock.

## 3 Feedstock productivity

Production rates (dry matter, DM) of more than 20 t $ha^{-1}$ $year^{-1}$ were reported for willow plantations in Sweden and Canada after 3-5 years of growth in small-scale fertilized experimental plantations (Labrecque and Teodorescu, 2003; Christersson, 1986). In commercial willow plantations, biomass production rates in the range of 5-11 t DM $ha^{-1}$ $year^{-1}$ or lower have been reported, with great variation between growers (Mola-Yudego and Aronsson, 2008; Dimitriou and Mola-Yudego, 2017; Aylott et al., 2008). Greater productivity rates are expected in the near future also for the commercial plantations, due to the more frequent use of improved willow varieties and better management of the plantations (Mola-Yudego, 2011).

In general, various plant-breeding methods are available to create more productive varieties of willow with improved quality traits for biofuel use. Owing to their shorter breeding histories, willows are less domesticated than most traditional agricultural crops; but due to a number of factors including short generation times, easy hybridization, clonal propagation, and large natural genetic variation, willow breeders have made great progress during the last decades (Karp et al., 2011). In addition, willows have a relatively small genome size and are closely related to *Populus*, a model genus for tree genetics and physiology (Ghelardini et al., 2014; Hanley et al., 2006; Berlin et al., 2010); and the genome of purple willow (*Salix purpurea*) has recently been sequenced (http://phytozome.jgi.doe.gov). Taken together, these factors and genomic tools can aid in developing a rich resource for genetic improvements of willows in the future.

As for most perennial crops, a major challenge in *Salix* breeding is the production of plant material with high productivity over many years, which stresses the need for specific breeding goals targeting traits important for cropping security (e.g. resistance to abiotic and biotic stresses) apart from productivity-related traits. Breeding programs for *Salix* were launched in many countries mainly of northern temperate regions (Kuzovkina et al., 2008; Stanton et al., 2014; Smart and Cameron, 2008), using both traditional crossings and breeding methods but increasingly also molecular methods and marker-assisted breeding. Willows can be used as raw material (biomass) for different biofuel supply chains, such as the production of wood chips for combustion or the production of gaseous or liquid fuels (e.g. biogas and bioethanol). Specific breeding objectives for biomass willows target not only productivity, but also growth habit (to adapt the crop to commercial harvesting machinery), water-use efficiency and drought tolerance, nitrogen-use efficiency, frost and disease resistance, and biofuel quality characteristics related to, for example, wood chip quality (Karp et al., 2011; Ronnberg-Wastljung et al., 2005; Weih et al., 2006; Berlin et al., 2014a; Fabio et al., 2017; Samils et al., 2011; Pawar et al., 2018).

Apart from the genetics of the plant material, the production system characteristics and management strongly affect both quantity and quality of the biomass produced in willow plantations (Verwijst et al., 2013; Adler et al., 2008). For example, the planting and weed control strategies have strong effects on the stand establishment, development, and biomass productivity (Albertsson et al., 2016; Welc et al., 2017; Edelfeldt et al., 2015; Bergkvist and Ledin, 1998; Nordh, 2005). Irrigation is currently not applied in most plantations of willow used for biofuel production, although the availability of water can be critical for achieving high productivity rates in some regions (Lindroth and Bath, 1999). High temperatures during the growing season are not critical for willow plantations, as long as the stands have good water supply (Bonosi et al., 2013). Nutrient fertilization increases willow productivity in many sites, but the effect of nutrient fertilization on productivity is strongly dependent on the plant material and site conditions (Aronsson et al., 2014; Weih and Nordh, 2005; Fabio and Smart, 2018b). Observational studies in differently composed tree stands suggest that also the composition of the plantation in terms of pure stands, consisting of single species or cultivars, or mixed stands can affect the biomass productivity (Liang et al., 2016; Verheyen et al., 2016). Currently, most commercial willow plantations grown for biomass production consist of single species or cultivars. So far, no clear evidence has been found for a strong positive effect of stand diversity on productivity in young willow plantations (Dillen et al., 2016; Hoeber et al., 2018), but it is possible that a (positive) diversity effect on productivity will evolve as plantations grow older. In addition, the specific species or cultivar composition in a willow plantation has been reported to significantly affect the biomass productivity and other growth factors in mixed willow stands, which suggests the possibility for designing specific willow mixtures to promote productivity as well as other characteristics beneficial for the ecological sustainability of willow plantations grown for biomass production (Hoeber et al., 2018; Baum et al., 2018).

## 4 Cropping security

Durable resistance toward pathogens and insects, tolerance toward abiotic factors such as frost and drought, and efficient nutrient acquisition and use are especially important in perennial crops such as willow SRC grown for biofuel production purposes, because these crops are supposed to grow for many years. One way to enhance crop protection and cropping security is to breed for increased resistance or tolerance toward the relevant abiotic and biotic factors. In willows grown for biofuel production, the ecophysiological mechanisms and genetic background of nutrient and water use as well as the resistance or tolerance toward frost, drought, and pathogens have been investigated (Tsarouhas et al., 2004; Pucholt et al., 2015; Samils et al., 2011; Weih et al., 2011, 2006; Pei et al., 2008; Bonosi et al., 2010; Ronnberg-Wastljung et al., 2005; Beyer et al., 2018), and many of the corresponding crop traits are frequently among the most prominent targets in willow breeding programs (Kuzovkina et al., 2008). Another way to increase crop protection and performance especially in perennial production systems is to utilize ecological (biological) processes and functions that support stress tolerance, resource-use efficiency and pest control, and thereby enhance the long-term sustainability of the production system (Weih et al., 2008; Rooney et al., 2009; Grossman et al., 2018), including the possibilities of biological control and integrated pest management (IPM) (Bjorkman et al., 2004; Moritz et al., 2017; Charles et al., 2014). Compared with the corresponding traditional management actions needed, utilization of these ecological processes usually requires more time to show effects, which provides opportunities particularly in perennial production systems with their considerably longer-time perspective compared to annual systems. The utilization of ecological processes to reduce the necessity of traditional management actions is therefore expected to require longer time periods (Fig. 5), but offers a viable alternative to enhance the long-term sustainability in perennial systems such as willow SRC. A major challenge in the utilization of ecological processes for improving long-term sustainability in these systems is to understand the complexity of the involved mechanisms. In willow grown for biofuel, approaches are being explored to utilize ecological information for improving crop protection toward harmful leaf beetles frequently damaging willow SRC (Bjorkman et al., 2004; Stenberg et al., 2010; Moritz et al., 2017), manipulating willow resource use, and enhancing pest resistance through the selection of appropriate soil microbe (e.g., mycorrhiza)-plant combinations (Fig. 6) (Rooney et al., 2009; Fransson et al., 2013; Baum et al., 2009b, 2018), and reducing damage to trees by pests and pathogens through the culture of more diverse (mixed) plantations (McCracken and Dawson, 1994, 1998; Muller et al., 2018). In addition, ecological research has addressed the induction of insect resistance in willows (Hoglund and Larsson, 2005; Hoglund et al., 2005),

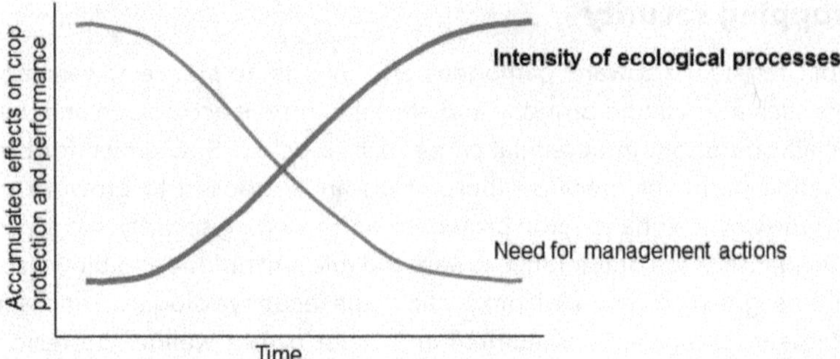


**Figure 5** Relationship between time and the intensity of ecological (biological) processes supporting crop protection and performance (e.g. protection against pests and pathogens, enhancement of microbial-driven nutrient acquisition) and the mirror image, indicating the need for management actions (e.g. pesticides, nutrient fertilizers) to achieve the same effects on crop protection and performance. Source: redrawn and modified after Weih et al. (2008).

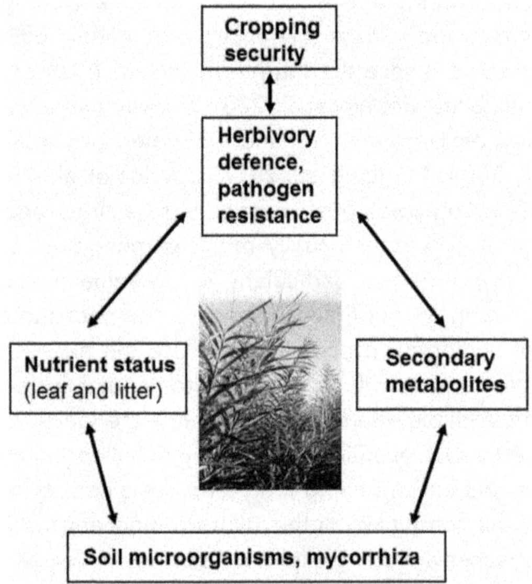


**Figure 6** Potential effects of soil organisms and mycorrhizal fungi on cropping security in willow plantations. Mycorrhizal fungi are an important integral component of the plant-soil system, forming symbiotic associations with willow roots and having a fundamental role in plant nutrition and protection against pathogens through effects on the foliar concentrations of, for example, secondary metabolites. Source: modified after Rooney et al. (2009). Photo: J. Svennås-Gillner, SLU, Sweden.

the factors affecting palatability of willow leaves for herbivores (Albrectsen et al., 2007, 2004; Glynn et al., 2004, 2007; Baum et al., 2009b), and willow leaf beetle behave with particular focus on the common management practice of willow plantations and the possibilities for biological control including IPM and plant breeding (Charles et al., 2014; Bjorkman et al., 2004; Dalin et al., 2004; Bjorkman and Ahrne, 2005; Bjorkman and Eklund, 2006; Stenberg et al., 2010). However, utilizing the knowledge on the biology and ecology of the organisms involved to improve cropping security of willow SRC in a predictable way remains a great challenge, as the relevant biological interactions are often highly context-dependent and greatly affected by abiotic factors, for example, temperature (Puentes et al., 2015; Fransson et al., 2013).

## 5 Biodiversity

On the one hand, the previous sections discussed some potentially beneficial effects of more diverse willow plantations (i.e. species or variety mixtures) on productivity and cropping security. On the other hand, studies on sustainability aspects of biomass production systems often point out negative effects of these production systems on biodiversity (landscape level) as a key concern (Desiree et al. 2014). In general, plantations of trees can have positive or negative effects on biodiversity at the landscape level, depending on location, planting design (e.g., single-tree or SRC), and management of the plantation–but also previous land use and the kinds of organisms considered (Hartley, 2002). Plantations of willow SRC grown on agricultural land have been reported to improve plant diversity at landscape level, in particular if the plantations are established instead of cultures of cereals and spruce or fallow ground in a homogeneous agricultural landscape (Baum et al., 2009c, 2012). The willow plantations can provide benefits to bird and small mammal populations and enhance their value for biodiversity, especially if multiple age classes of willows are maintained in the landscape simultaneously (Berg, 2002; Campbell et al., 2012). Also, high number and diversity of beneficial insects has been found in willow plantations when compared to conventional agricultural fields (cereals, sugar beet, and potato) (Jaworska, 2010). Comprehensive information regarding the effects of willow SRC on the biodiversity of many animals is still lacking, but it has been concluded that (1) the diversity of breeding birds is higher than in agricultural cropland, but lower than in forest ecosystems (Berg, 2002; Campbell et al., 2012); (2) ground beetle diversity is higher in arable fields than in willow SRC (Jaworska, 2010); (3) willow SRC plantations accommodate both a greater diversity and higher abundance in most animal groups compared to poplar plantations (Schulz et al., 2009); and (4) the cultivation of willow SRC can lead to an increase in animal diversity in homogeneous agricultural landscapes, but to adverse effects in landscapes with high conservational value (Schulz et al., 2009). Little information is available on the soil fungal and bacterial responses to the conversion of conventional agricultural land to willow SRC, reporting enrichment of fungal communities (including mycorrhizal fungi) but indifferent responses of bacterial communities (Xue et al., 2016). In contrast to most conventional agricultural crops, willow roots can be colonized by ectomycorrhizal fungi, leading to the introduction of these fungi into arable soils and to considerable changes in the soil microbial colonization and activity when conventional agricultural land is converted to willow SRC. The abundance and diversity of mycorrhizal fungi associated with willows was found to be greatly affected by the previous land use, management (e.g. nutrient fertilization), the plant material, and the age of the willow plants (Hrynkiewicz et al., 2010, 2012; Baum et al., 2002, 2009a). Compared to conventional agricultural land, the non-tillage management and the high litter supply in willow SRC can increase the abundance of earthworms but decrease the abundance of certain ground-living beetles (Carabidae) (Baum et al., 2009a). In summary, the effects of planting willow SRC after conventional agricultural crops on landscape biodiversity are frequently positive, especially in open agricultural landscapes and when maintaining high structural diversity (e.g. multiple age classes), but effects are strongly context-dependent.

## 6 Nutrient uptake and use

Reduced resource depletion is an important element in sustainable land use for biofuel production purposes (Higman et al., 2005; Ra et al., 2012), and the availability of mineral nutrients and particularly nitrogen (N) is a critical factor in willow production for biofuel, because the manufacture of commercial fertilizers consumes much fossil-based energy and thereby contributes to global warming. Biomass productivity per unit plant-internal amount of N is considerably higher in willows compared to conventional agricultural crops such as wheat (Weih et al., 2018). Increased productivity at maintained nutrient accumulation in the crop would be one way to improve crop nutrient use, and some of the key factors and mechanisms determining the productivity per unit plant-internal N under different environmental conditions (e.g., drought) and using different plant material have been identified for willows used for biomass production (Weih et al., 2011; Weih and Nordh, 2002; Weih, 2009). Another way of increasing productivity is the use of nutrient fertilizers, and willows used for biomass production greatly vary in their growth response to different nutrient supply (Weih and Nordh, 2005; Fabio and Smart, 2018a). Thus, enhanced biomass production through increased nutrient uptake rates is a prominent target in improvement programs for biomass crops, but is not in line with the sustainability aim of reduced resource depletion, because higher productivity is usually associated with increased N absorption from the soil (Ra et al., 2012). In general, nutrient fertilization greatly enhances crop yields, but decreases the productivity per unit of plant-internal amount of N (Weih et al., 2018), reflecting the general pattern of more than proportional resource demand when crop yields are increased through fertilization (Tilman et al., 2002). In an ecological sustainability context, a major breeding focus for willow cultivars to be grown for biofuel production could be enhanced energy output per N resource depletion, and the identification of heritable traits affecting it (Weih et al., 2014). This ratio is expected to be influenced by various crop characteristics, for example, the productivity per unit plant-internal N and quality traits of the harvested biomass. The productivity per unit plant-internal N is expected to be enhanced through the traits optimizing canopy architecture and photosynthesis, for which genetic variation and breeding potential has been documented in willows (Weih and Ronnberg-Wastljung, 2007; Fabio and Smart, 2018a). Additional breeding opportunities relevant for enhanced energy output per N resource depletion arise from the genetic variation in heating value and biomass quality traits documented for willows (Fabio et al., 2017; Pawar et al., 2018). Compared to a traditional focus of crop breeding research on biomass yields and qualities, a resource-oriented approach adopts a more holistic approach including resource (here nutrients) aspects that can be helpful in the evaluation of biofuel crops in an ecological sustainability context. Such an approach could also consider the energy produced in relation to the fate of the targeted resource. For example, the nutrient resources to generate feedstock for biogas production can easily be reused for the cultivation of new feedstock, resulting in enhanced sustainability of the production system compared to a system without the immediate reuse of the nutrient resources.

## 7 Carbon accumulation and sequestration

Considering the frequently environmental motivation of willow grown for biofuel production, an important target for the further development of bioenergy crops should be increased carbon transfer from atmosphere to soil and enhanced carbon accumulation in soil. Increased carbon accumulation has been documented in willow plantations grown on formerly arable soils, while indifferent or positive effects on carbon accumulation were found when willow SRC was planted on former grassland sites (Kahle et al., 2005; Harris et al., 2015, 2017; Dimitriou et al., 2012b). Also, carbon balance models targeting the carbon fluxes between soil, biomass, and atmosphere in willow SRC indicated that these plantations act as carbon sinks and therefore have a mitigating effect on climate change, especially when plant material and management conditions are applied that support high biomass yields (Hammar et al., 2014). An important factor affecting the total carbon fluxes and the soil carbon accumulation in a willow stand is the previous land use (Grogan and Matthews, 2002). Also, the choice of the plant material, for example, willow genotype can affect below-ground biomass allocation in the plant and thereby soil carbon accumulation, which opens possibilities for plant breeding toward increased carbon accumulation under willow plantations grown for biofuel production (Cunniff et al., 2015; Gregory et al., 2018). Apart from the quantity of accumulated soil carbon, its specific composition is also important and determines the longevity of the carbon pool in the soil and thus the sustainability of any carbon accumulation seen in the soil under biomass crops (Baum et al., 2013). The formation of root mycorrhizal fungi can also enhance the amount of carbon allocated below-ground under willow (Jones et al., 1991), and these root-associated fungi can affect several host plant traits relevant for soil carbon accumulation both directly through the roots and indirectly through the influences on leaf and litter chemistry (Rooney et al., 2009). However, the interaction between the relevant willow plant traits and the fungi appears difficult to predict due to the complex nature of the corresponding interactions (Fransson et al., 2013).

## 8 Ecosystem services and environmental impact

Ecosystem services are the goods and services provided by an ecosystem to society, including watershed services, nutrient cycling, waste management, carbon storage and effects on climate change, scenic landscapes, and biodiversity and wildlife habitat. Many of these services were already discussed in the previous sections of this chapter regarding willows used for biofuel production, and are also reviewed elsewhere (Isebrands et al., 2014). This section highlights some environmental applications that are relevant in the context of the sustainable use of willows used for biofuel production, and summarizes the reported knowledge on the environmental impact of willow SRC including some results from life cycle assessment (LCA).

Productive willow plantations take up large quantities of water and nutrients, which has initiated various applications in which these plantations are used as recipients for municipal wastewater and industrial sludge along with the simultaneous production of biomass, sometimes referred to as multifunctional biomass plantations (Aronsson and Perttu, 2001; Labrecque and Teodorescu, 2003). A major challenge occurs with those applications in regions with extended dormant periods during winter, where only small quantities of nutrients are taken up by the willow plants; in those regions, wastewater can be stored in ponds or lagoons during the winter period for subsequent application to the plantations during the growing season (Dimitriou and Aronsson, 2005).

A large-scale shift from, for example, conventional agricultural crops to willow SRC is likely to impact various environmental issues, although there are many uncertainties with respect to the specific impacts in a given context, because willow SRC still is a new production system for most regions in which it could be grown in the future. Thus, results so far indicate many environmental benefits, but also negative impacts on the environment due to willow SRC implementation, and the effects on the environment strongly depend on the existing or previous land use, the plantation scale, the plant material used, and the management practices applied (Weih and Dimitriou, 2012; Langeveld et al., 2012; Bacenetti et al., 2016). Apart from the effects of willow SRC on soil carbon accumulation and biodiversity summarized previously, willow SRC has been reported to decrease leaching of nitrate to the groundwater, but increase phosphorus leaching compared to conventional agricultural crops (Dimitriou et al., 2012a; Schmidt-Walter and Lamersdorf, 2012), and decrease the soil concentration of cadmium (Cd), a hazardous trace element for human health in the food chain (Dimitriou et al., 2012b).

LCA revealed a considerably higher production efficiency and environmental performance of willow biomass used for direct combustion in combined heat and power plants (CHP) compared to fossil fuel-based power plants (Buonocore et al., 2012). Compared to other feedstock used for biofuel production, willow has a high energy output to input ratio and a low carbon footprint (Parajuli et al., 2017). In willow SRC used for biofuel, the most important factors affecting the energy efficiency are the application of nutrient fertilizers (affecting yield level), the harvest, and the transport of the harvested product (Hammar et al., 2017). The application of nutrient fertilizers to willow plantations enhances the productivity, but also the negative environmental impacts related to, for example, eutrophication and toxicity potentials. Taken together, it has been shown that the higher biomass productivity achieved with nutrient fertilization results in a better overall environmental profile in terms of net energy yield (i.e. net quantity of energy in MJ generated per land area) and global warming potential compared to unfertilized plantations (Gonzalez-Garcia et al., 2012). However, there is a trade-off between achieving a high net energy yield and achieving a high energy ratio (Nordborg et al., 2018), and a common notion in many LCA and environmental impact studies is the strong context-dependency of the environmental impacts associated with the plantation of willow SRC for biofuel.

## 9 Case study

In Sweden, willows have been grown for biofuel purpose for several decades (Christersson et al., 1993), and Sweden has highly ambitious climate goals aiming at zero net greenhouse gas emissions by 2045 (e.g. https://www.newscientist.com/article/2138008). Moving from a fossil-based to a bio-based society with reduced greenhouse gas emissions requires an increased use of multiple sources of renewable energy, and the sustainable use of willow for biofuel production provides a viable alternative for contributing to the transition from a fossil-based to a bio-based society. For willows to play a significant role in this development, new willow varieties are needed that can be used for the production of different biofuels and contribute to sustainable development in terms of, for example, enhanced productivity and cropping security, increased resource-use efficiency, and a large carbon sequestration potential.

In perennial plants such as willows grown in seasonal environments, the timing of the critical stages for growth initiation in spring (e.g. bud-burst and advancement of leaf unfolding) and growth termination in autumn (e.g. height growth cessation and leaf abscission), that is, leaf phenology, is an important driver for biomass productivity. Spring phenology determines the start of the growth period, but earlier bud-burst may increasingly predispose plants to greater risk of frost damage. Autumn phenology is closely linked to frost-hardening, and a correlation between frost resistance and height growth cessation has been reported in willow (Ogren, 1999). However, many willows continue to be photosynthetically active and accumulate biomass after height growth cessation and bud set (Bollmark et al., 1999). Thus, a delayed growth cessation and/or autumn leaf fall could go along with increased shoot production, but light and temperature conditions are rapidly deteriorating toward late autumn especially in high-latitude environments, and there should be a critical period in autumn after which delayed growth cessation and/or leaf fall is not any longer associated with enhanced productivity. Also for biomass willows, the bud and leaf phenologies have been pointed out as promising targets for the breeding (Lennartsson and Ogren, 2004; Ronnberg-Wastljung, 2001). However, there is still uncertainty in how the variation in phenology affects the biomass production, and also about the genetic background of those phenology characteristics that most strongly affect biomass production. Due to these uncertainties, various commercial willow varieties have been examined to investigate the effects of willow cultivar and environment on spring and autumn phenologies as well as the relationships between phenology and biomass production; the genetic background of the corresponding phenology traits was analyzed in a large breeding population. The duration of the leafy period, that is, the period between bud-burst and leaf abscission, and bud-burst date were found to strongly affect the productivity of the willows; the timing of bud-burst and leaf abscission was shown to be more important for willow biomass production than the timing of height growth cessation (Weih, 2009) (Fig. 7). Subsequent studies investigated the genetic background of the corresponding phenology characteristics in a willow breeding population by assessing genetic and phenotypic variation in bud-burst, growth cessation, and leaf abscission in different years and environments, and identified various genetic markers for these traits that will speed up the breeding of biomass willows in the future (Ghelardini et al., 2014; Berlin et al., 2017). In a similar approach, critical characteristics for willow resource (nutrients and water) economy were first analyzed at the phenotype level in a breeding population (Weih et al., 2011), and the genetic background of the same characteristics was subsequently evaluated and the corresponding genetic markers were identified

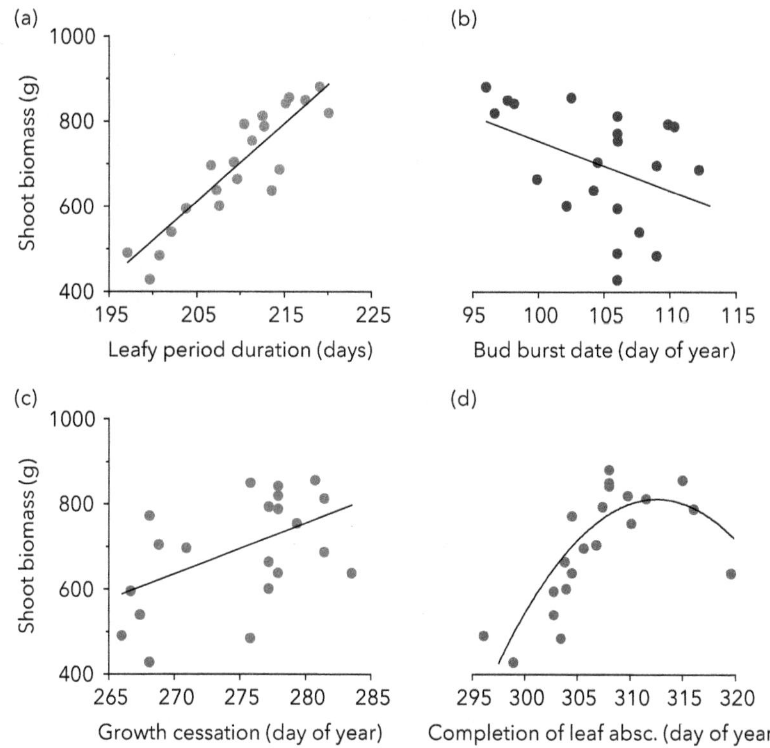


**Figure 7** The shoot biomass production of willow as related to the duration of period in which the trees had leaves (a), bud-burst date (b), timing of growth cessation (c), and leaf abscission (d). The timing of bud-burst and leaf abscission are better correlated to biomass production than the timing of height growth cessation. Source: redrawn from Weih (2009).

to support rapid breeding for sustainable willow production and use (Berlin et al., 2014a). Apart from productivity characteristics focusing on the amount of biomass produced, recent research targets increasingly also biomass quality and different routes for biofuel use from willows, for example, the identification of genetic markers for different wood chemical traits, information that is important for willow breeding toward different uses of its biomass (Pawar et al., 2018). The knowledge and tools that have been built up within the framework of the above research are excellent resources for a continued plant breeding of willow and in addition also make it possible to develop willow into a model plant for studying breeding strategies in woody plant species.

When evaluating the sustainable use of willow for biofuel production, the breeding issues discussed so far need to be connected to ecological functions and ecological sustainability. In general, plant breeding is strictly trait-based and usually targets relatively short-term events occurring in more or less

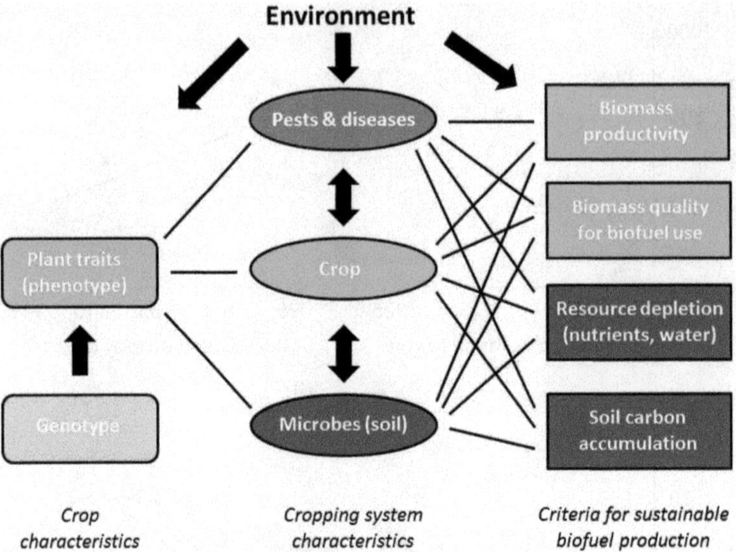


**Figure 8** Conceptual overview of the relationships between the crop characteristics targeted in plant breeding; the cropping system characteristics represented by the crop, its pests and diseases, and the soil microbial life; and some criteria for the evaluation of sustainable biofuel production and use. Different colors indicate genotype level (orange), plant/crop level (green), above-ground biotic interaction (blue), and below-ground biotic interaction (maroon). Source: modified from Weih et al. (2014).

simple systems, whereas any ecological sustainability evaluation requires the integration of longer-term ecological processes occurring in more complex systems (Weih et al., 2008). Therefore, a trait-to-ecosystem perspective should be integrated into the breeding research for the sustainable use of plants as feedstock for biofuel production. In such a perspective, the cascading effects of plant characteristics targeted in breeding research on the ecosystem processes particularly important in the evaluation of the ecological sustainability of biofuel crops should be considered (Weih et al., 2014) (Fig. 8). Adopting such an approach, some willow genotypes have been found to positively affect community productivity, resource (nitrogen) efficiency, and soil carbon accumulation while others seem to have negative effects on the same functions (Hoeber et al., 2017, 2018; Gregory et al., 2018), but the corresponding specific plant traits behind those effects have yet to be identified.

## 10 Summary and future trends

Intensively managed willow plantations are gaining interest, due mainly to their efficient and sustainable land use in combination with an increasing demand for biofuel resources. In many countries, willows have been used commercially as biofuel in the form of wood chips for direct combustion in heat and power plants, but there is an increasing interest in using biomass from willows as raw material for conversion into biofuels such as biogas and bioethanol. The establishment of willow plantations for biofuel production purposes is often motivated in terms of beneficial effects on the environment and sustainable development. A major challenge is striking the balance between long-term sustainability and reaching short-term crop productivity goals. Plant breeding is an efficient tool to create more productive varieties with improved quality traits for biofuel use, and major breeding progress has been achieved in willows during the last decades using both traditional breeding methods and increasingly also molecular methods and marker-assisted breeding. Major challenge in willow breeding is the production of plant material with high productivity over many years, with focus on specific traits important for cropping security (e.g. resistance to abiotic and biotic stresses) apart from purely productivity-related traits including resource-use (water and nutrients) efficiency and biofuel quality. Besides genetics and breeding, the production system characteristics and management strongly affect the quantity and quality of the biomass produced. Productivity is affected by the composition of the plantation in terms of pure stands consisting of single species or cultivars vs. mixed stands, involving the possibility for designing specific willow mixtures to promote productivity as well as other characteristics beneficial for the ecological sustainability of willow biofuel plantations. Durable resistance toward pathogens and insects, tolerance toward abiotic factors such as frost and drought, and efficient nutrient acquisition and use are especially important in willows grown for biofuel production, because these perennial crops are supposed to grow for many years. An attractive alternative to achieve durable resistance is the utilization of ecological processes and functions that support stress tolerance, resource-use efficiency, and pest control and thereby enhance the long-term ecological sustainability of the production system; a major challenge in the utilization of ecological processes is to understand the complexity of the involved mechanisms. Studies on ecological sustainability aspects of biofuel production systems sometimes point out negative effects on landscape-level biodiversity. For willows, the effects on landscape biodiversity are frequently reported positive, although comprehensive information regarding the effects of willow plantations on the biodiversity of many organisms is still lacking. The reduced depletion of resources such as nutrient fertilizer is an important element in sustainable land use for biofuel production, and a major breeding focus could be enhanced energy output per nutrient depletion, and the identification of heritable traits affecting it. Increased carbon transfer from atmosphere to soil and enhanced soil carbon accumulation are other important goals in the sustainable use of willow for biofuel production, since these factors have been shown to be especially important for the total climate effects in a life cycle perspective. Willow plantations have been reported to frequently act as carbon sinks, but soil carbon accumulation is affected by various factors including site history, choice of plant material, and soil microbial life. Willow plantations are also used as recipients for municipal wastewater and industrial sludge along with the simultaneous production of biomass and can thus provide multiple ecosystem services, with positive environmental impacts. A large-scale shift from, for example, conventional agricultural crops to willow plantations is likely to impact various environmental issues, although there are many uncertainties with respect to the specific impacts in a given context. Compared to other feedstock used for biofuel production, willow has a high energy output to input ratio and a low carbon footprint, and research results indicate many environmental benefits, but also negative impacts on the environment.

Intensively grown willow plantations have the potential to become key players in the sustainable production of woody biomass to be used as raw material for various biofuel supply chains, and willows are thus interesting as feedstock not only for direct combustion in heat and power plants, but also in various biorefinery conversion processes (Sassner et al., 2008b; Horn et al., 2011; Serapiglia et al., 2013b). Development and breeding of willow plant material better suited for the use as biofuel crops compared to the currently available material requires an improved understanding of the structure and chemistry of willow wood and knowledge about the genetic variation in, and control over, the relevant wood traits. The relevant wood traits include the characteristics facilitating, for example, improved lignocellulose disruption and subsequent enzymatic saccharification, and their interactions with different microbiological and thermochemical biofuel production systems. Molecular breeding approaches using marker-assisted selection in plant breeding are promising for wood characteristics, as the relevant traits are time consuming and costly to measure. Thus, the development of genetic markers for key genes controlling wood traits would greatly facilitate the early selection and efficient breeding of willows for biofuel use, and thus increase the potential genetic gain. Almost half a century of pre-breeding research and commercial breeding of willow in Sweden and elsewhere gives access to a wide range of well-defined willow plant material to be used for future pre-breeding research, and there are several established research populations harboring genetic resources for future breeding programs. Especially in the context of the sustainability goals important for the use of willows for biofuel, holistic breeding approaches considering the ecological processes and functions supporting long-term cropping security (i.e. resistance and/or tolerance to pests and diseases, reduced depletion of water and nutrient resources) probably will be important research subjects in the future, although the complexity of the involved biological and ecological processes will be a major challenge for the predictable utilization of ecological interactions in terms of, for example, the biological control of pests including IPM (Charles et al., 2014; Weih et al., 2008). Especially in light of the future development toward the use of willow for diverse biofuel outputs, more research is needed on the effects of carbon-rich waste products from various biorefinery processes returned to the soil on the below-ground processes driving carbon cycling and sequestration; the latter is one of the ecosystem services of great public concern in relation to climate change and maintenance of soil fertility. For example, the effects of plant material (e.g. variety) identity, and plantation diversity (e.g., defined variety mixtures) on biofuel productivity and quality, cropping security of the plantation, and below-ground carbon and nutrient cycling need to be evaluated in the future to ensure a sustainable use of willow for different biofuel production chains. In this context, LCA has emerged as a powerful tool for quantifying and improving the environmental impacts of biofuel production in a sustainability perspective (Ahlgren et al., 2012), and supported by recent progress in methodology developments including dynamic modeling of carbon dynamics and global temperature increase using response functions (Ericsson et al., 2013; Hammar et al., 2014). A future trend is thus to combine, in an integrated approach, the research on willow genetics and breeding with investigations on biological and chemical wood conversion processes, ecological processes, and biological control including IPM, carbon and nutrient cycling, and environmental impact by using, for example, LCA methodology (Fig. 9).

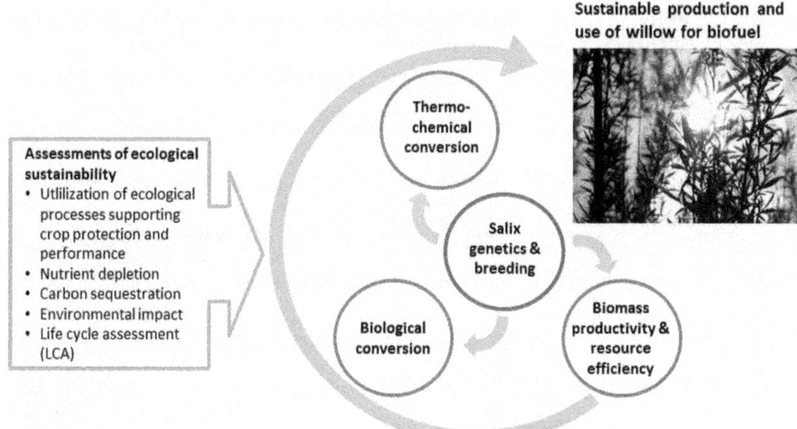


**Figure 9** Outline of a research approach integrating willow genetics and breeding with investigations on biological and chemical wood conversion processes and the assessment of ecological sustainability criteria related to the utilization of ecological processes for, for example, biological control, carbon and nutrient cycling, and environmental impact by using, for example, LCA methodology. Photo: V. Wrange, SLU, Sweden.

## 11 Acknowledgements

The compilation of this chapter was partly funded by the The Swedish Research Council for Environment, Agricultural Sciences and Spatial Planning (Formas, project no. 942-2016-31).

## 12 Where to look for further information

Updated information on all aspects of willows grown for biofuel production and other purposes is found at the web page of the International Poplar Commission (IPC) at: www.fao.org/forestry/ipc. Regular conferences addressing current aspects of willow (and poplar) production and use for various purposes are organized by the IPC (information at the above website) and by the International Union of Forest Research Organizations (IUFRO), unit 2.08.04–Poplars and Willows; see further information at: www.iufro.org/science/divisions/division-2/20000/20800/20804. Several reviews have been published regarding the sustainable production and use of willow (and other plants) for biofuel, for example, Volk et al. (2004, 2016), Verwijst et al. (2013), Karp and Shield (2008), Amichev et al. (2014), and Weih (2004). Overviews on various willow taxonomy, genetics, domestication, and breeding aspects are given by Gullberg (1993), Larsson (1997), Kuzovkina et al. (2008), Dickmann and Kuzovkina (2014), Stanton et al. (2014), Smart and Cameron (2008), Karp et al. (2011), Hanley and Karp (2014), Berlin et al. (2014b), Hallingback et al. (2016), and Ronnberg-Wastljung and Gullberg (1999), while management aspects are reviewed by Verwijst et al. (2013) and Stanturf and van Oosten (2014) and documented in manuals for growers published for different regional contexts at various publicly available web pages. Basic wood quality aspects of willows are reported by Ledin (1996), while wood properties, processing, and utilization of willow biomass for energetic use is reviewed by Balatinecz et al. (2014) and various biofuel quality aspects for different willow supply chains are subjects of the more specialized papers by Jirjis (2005), Serapiglia et al. (2013b), Pawar et al. (2018), Whittaker et al. (2018), and Krzyzaniak et al. (2016a). Willow environmental applications, ecosystem services, and environmental impact assessments including biodiversity are reviewed and summarized in several book chapters and/or special issues of journals, for example, Isebrands et al. (2014), Dimitriou and Aronsson (2005), Dimitriou et al. (2009), Weih and Dimitriou (2012), Weih and Nordh (2009), and Bacenetti et al. (2016). The ecology and physiology of willows, including the responses to abiotic stresses, have been reviewed by Mitchell et al. (1992), Richardson et al. (2014), and Marron et al. (2014), whereas specific reviews on willow diseases and pests including IPM where published by Ostry et al. (2014) and Charles et al. (2014). The mathematical modeling of biomass production in willow plantations grown under different environmental conditions has been addressed (Wang et al., 2015; Sannervik et al., 2006; Cerasuolo et al., 2016; Philippot, 1996), as well as different methodologies of LCA to simulate environmental and climate impacts for different willow biofuel supply chains (Heller et al., 2003; Gonzalez-Garcia et al., 2012; Buonocore et al., 2012; Budsberg et al., 2012; Krzyzaniak et al., 2016b; Ericsson et al., 2017; Hammar et al., 2017).

## 13 References


Adler, A., Dimitriou, I., Aronsson, P., Verwijst, T. and Weih, M. 2008. Wood fuel quality of two *Salix viminalis* stands fertilised with sludge, ash and sludge-ash mixtures. *Biomass and Bioenergy* 32(10), 914–25. doi:10.1016/j.biombioe.2008.01.013.

Ahlgren, S., Roos, E., Lucia, L. D., Sundberg, C. and Hansson, P. A. 2012. EU sustainability criteria for biofuels: uncertainties in GHG emissions from cultivation. *Biofuels* 3(4), 399–411. doi:10.4155/bfs.12.33.

Albertsson, J., Verwijst, T., Rosenqvist, H., Hansson, D., Bertholdsson, N. O. and Ahman, I. 2016. Effects of mechanical weed control or cover crop on the growth and economic viability of two short-rotation willow cultivars. *Biomass and Bioenergy* 91, 296–305. doi:10.1016/j.biombioe.2016.05.030.

Albrectsen, B. R., Gardfjell, H., M. Orians, C., Murray, B. and S. Fritz, R. 2004. Slugs, willow seedlings and nutrient fertilization: intrinsic vigor inversely affects palatability. *Oikos* 105(2), 268–78. doi:10.1111/j.0030-1299.2004.12892.x.

Albrectsen, B. R., Gutierrez, L., Fritz, R. S., Fritz, R. D. and Orians, C. M. 2007. Does the differential seedling mortality caused by slugs alter the foliar traits and subsequent susceptibility of hybrid willows to a generalist herbivore? *Ecological Entomology* 32, 070130195410003–???. doi:10.1111/j.1365-2311.2006.00860.x.

Alexandropoulou, M., Antonopoulou, G., Ntaikou, I. and Lyberatos, G. 2017. Fungal pretreatment of willow sawdust with *Abortiporus biennis* for anaerobic digestion: impact of an external nitrogen source. *Sustainability* 9(1), 14. doi:10.3390/su9010130.

Amichev, B. Y., Hangs, R. D., Konecsni, S. M., Stadnyk, C. N., Volk, T. A., Belanger, N., Vujanovic, V., Schoenau, J. J., Moukoumi, J. and Van Rees, K. C. J. 2014. Willow short-rotation production systems in Canada and northern United States: a review. *Soil Science Society of America Journal* 78(S1), S168–82. doi:10.2136/sssaj2013.08.0368nafsc.

Aronsson, P. and Perttu, K. 2001. Willow vegetation filters for wastewater treatment and soil remediation combined with biomass production. *The Forestry Chronicle* 77(2), 293–9. doi:10.5558/tfc77293-2.

Aronsson, P., Rosenqvist, H. and Dimitriou, I. 2014. Impact of nitrogen fertilization to short-rotation willow coppice plantations grown in Sweden on yield and economy. *Bioenergy Research* 7(3), 993–1001. doi:10.1007/s12155-014-9435-7.

Aylott, M. J., Casella, E., Tubby, I., Street, N. R., Smith, P. and Taylor, G. 2008. Yield and spatial supply of bioenergy poplar and willow short-rotation coppice in the UK. *The New Phytologist* 178(2), 358–70. doi:10.1111/j.1469-8137.2008.02396.x.

Bacenetti, J., Bergante, S., Facciotto, G. and Fiala, M. 2016. Woody biofuel production from short rotation coppice in Italy: environmental-impact assessment of different species and crop management. *Biomass and Bioenergy* 94, 209–19. doi:10.1016/j.biombioe.2016.09.002.

Balatinecz, J., Mertens, P., De Boever, L., Hua, Y. K., Jin, J. W. and Van Acker, J. 2014. Properties, processing and utilization. In: Isebrands, J. G. and Richardson, J. (Eds), *Poplars and Willows: Trees for Society and the Environment*. CABI Publishing, Wallingford.

Baum, C., Weih, M., Verwijst, T. and Makeschin, F. 2002. The effects of nitrogen fertilization and soil properties on mycorrhizal formation of *Salix viminalis*. *Forest Ecology and Management* 160(1–3), 35–43. doi:10.1016/S0378-1127(01)00470-4.

Baum, C., Leinweber, P., Weih, M., Lamersdorf, N. and Dimitriou, I. 2009a. Effects of short rotation coppice with willows and poplar on soil ecology. *Landbauforschung Volkenrode* 59, 183–96.

Baum, C., Toljander, Y. K., Eckhardt, K.-U. and Weih, M. 2009b. The significance of host-fungus combinations in ectomycorrhizal symbioses for the chemical quality of willow foliage. *Plant and Soil* 323(1–2), 213–24. doi:10.1007/s11104-009-9928-x.

Baum, S., Weih, M., Busch, G., Kroiher, F. and Bolte, A. 2009c. The impact of Short Rotation Coppice plantations on phytodiversity. *Landbauforschung Volkenrode* 59, 163–70.

Baum, S., Bolte, A. and Weih, M. 2012. High value of short rotation coppice plantations for phytodiversity in rural landscapes. *Global Change Biology Bioenergy* 4(6), 728–38. doi:10.1111/j.1757-1707.2012.01162.x.

Baum, C., Eckhardt, K. U., Hahn, J., Weih, M., Dimitriou, I. and Leinweber, P. 2013. Impact of poplar on soil organic matter quality and microbial communities in arable soils. *Plant, Soil and Environment* 59(3), 95–100. doi:10.17221/548/2012-PSE.

Baum, C., Hrynkiewicz, K., Szymanska, S., Vitow, N., Hoeber, S., Fransson, P. M. A. and Weih, M. 2018. Mixture of *Salix* genotypes promotes root colonization with dark septate endophytes and changes P cycling in the mycorrhizosphere. *Frontiers in Microbiology* 9, 1012. doi:10.3389/fmicb.2018.01012.

Berg, Å 2002. Breeding birds in short-rotation coppices on farmland in central Sweden - the importance of *Salix* height and adjacent habitats. *Agriculture, Ecosystems and Environment* 90(3), 265–76. doi:10.1016/S0167-8809(01)00212-2.

Bergkvist, P. and Ledin, S. 1998. Stem biomass yields at different planting designs and spacings in willow coppice systems. *Biomass and Bioenergy* 14(2), 149–56. doi:10.1016/S0961-9534(97)10021-6.

Berlin, S., Lagercrantz, U., Von Arnold, S., Ost, T. and Ronnberg-Wastljung, A. C. 2010. High-density linkage mapping and evolution of paralogs and orthologs in *Salix* and *Populus*. *BMC Genomics* 11, 129. doi:10.1186/1471-2164-11-129.

Berlin, S., Ghelardini, L., Bonosi, L., Weih, M. and Ronnberg-Wastljung, A. C. 2014a. QTL mapping of biomass and nitrogen economy traits in willows (*Salix* spp.) grown under contrasting water and nutrient conditions. *Molecular Breeding* 34(4), 1987–2003. doi:10.1007/s11032-014-0157-5.

Berlin, S., Trybush, S. O., Fogelqvist, J., Gyllenstrand, N., Hallingback, H. R., Ahman, I., Nordh, N. E., Shield, I., Powers, S. J., Weih, M., et al. 2014b. Genetic diversity, population structure and phenotypic variation in European *Salix viminalis* L. (Salicaceae). *Tree Genetics and Genomes* 10(6), 1595–610. doi:10.1007/s11295-014-0782-5.

Berlin, S., Hallingback, H. R., Beyer, F., Nordh, N. E., Weih, M. and Ronnberg-Wastljung, A. C. 2017. Genetics of phenotypic plasticity and biomass traits in hybrid willows across contrasting environments and years. *Annals of Botany* 120(1), 87–100. doi:10.1093/aob/mcx029.

Beyer, F., Jäck, O., Manzoni, S. and Weih, M. 2018. Relationship between foliar δ13C and sapwood area indicates different water use patterns across 236 *Salix* genotypes. *Trees-Structure and Function* 32(6), 1737–50, doi:10.1007/s00468-018-1747-3.

Bhagia, S., Muchero, W., Kumar, R., Tuskan, G. A. and Wyman, C. E. 2016. Natural genetic variability reduces recalcitrance in poplar. *Biotechnology for Biofuels* 9, 106. doi:10.1186/s13068-016-0521-2.

Bjorkman, C. and Ahrne, K. 2005. Influence of leaf trichome density on the efficiency of two polyphagous insect predators. *Entomologia Experimentalis et Applicata* 115(1), 179–86. doi:10.1111/j.1570-7458.2005.00284.x.

Bjorkman, C. and Eklund, K. 2006. Factors affecting willow leaf beetles (*Phratora vulgatissima*) when selecting overwintering sites. *Agricultural and Forest Entomology* 8(2), 97–101. doi:10.1111/j.1461-9555.2006.00288.x.

Bjorkman, C., Bommarco, R., Eklund, K. and Hoglund, S. 2004. Harvesting disrupts biological control of herbivores in a short-rotation coppice system. *Ecological Applications* 14(6), 1624–33. doi:10.1890/03-5341.

Bollmark, L., Sennerbyforsse, L. and Ericsson, T. 1999. Seasonal dynamics and effects of nitrogen supply rate on nitrogen and carbohydrate reserves in cutting-derived *Salix viminalis* plants. *Canadian Journal of Forest Research* 29(1), 85–94. doi:10.1139/x98-183.

Bonosi, L., Ghelardini, L. and Weih, M. 2010. Growth responses of 15 Salix genotypes to temporary water stress are different from the responses to permanent water shortage. *Trees-Structure and Function* 24(5), 843–54. doi:10.1007/s00468-010-0454-5.

Bonosi, L., Ghelardini, L. and Weih, M. 2013. Towards making willows potential bio-resources in the South: northern *Salix* hybrids can cope with warm and dry climate when irrigated. *Biomass and Bioenergy* 51, 136–44. doi:10.1016/j.biombioe.2013.01.009.

Budsberg, E., Rastogi, M., Puettmann, M. E., Caputo, J., Balogh, S., Volk, T. A., Gustafson, R. and Johnson, L. 2012. Life-cycle assessment for the production of bioethanol from willow biomass crops via biochemical conversion. *Forest Products Journal* 62(4), 305–13. doi:10.13073/FPJ-D-12-00022.1.

Buonocore, E., Franzese, P. P. and Ulgiati, S. 2012. Assessing the environmental performance and sustainability of bioenergy production in Sweden: a life cycle assessment perspective. *Energy* 37(1), 69–78. doi:10.1016/j.energy.2011.07.032.

Campbell, S. P., Frair, J. L., Gibbs, J. P. and Volk, T. A. 2012. Use of short-rotation coppice willow crops by birds and small mammals in central New York. *Biomass and Bioenergy* 47, 342–53. doi:10.1016/j.biombioe.2012.09.026.

Cerasuolo, M., Richter, G. M., Richard, B., Cunniff, J., Girbau, S., Shield, I., Purdy, S. and Karp, A. 2016. Development of a sink-source interaction model for the growth of short-rotation coppice willow and in silico exploration of genotype x environment effects. *Journal of Experimental Botany* 67(3), 961–77. doi:10.1093/jxb/erv507.

Charles, J. G., Nef, L., Allegro, G., Collins, C. M., Delplanque, A., Gimenez, R., Hoglund, S., Jiafu, H., Larsson, S., Luo, Y., et al. 2014. Insect and other pests of poplars and willows. In: Isebrands, J. G. and Richardson, J. (Eds), *Poplars and Willows: Trees for Society and the Environment*. CABI Publishing, Wallingford.

Christersson, L. 1986. High technology biomass production by *Salix* clones on a sandy soil in southern Sweden. *Tree Physiology* 2(1_2_3), 261–72. doi:10.1093/treephys/2.1-2-3.261.

Christersson, L., Sennerby-Forsse, L. and Zsuffa, L. 1993. The role and significance of woody biomass plantations in Swedish agriculture. *The Forestry Chronicle* 69(6), 687–93. doi:10.5558/tfc69687-6.

Cunniff, J., Purdy, S. J., Barraclough, T. J. P., Castle, M., Maddison, A. L., Jones, L. E., Shield, I. F., Gregory, A. S. and Karp, A. 2015. High yielding biomass genotypes of willow (*Salix* spp.) show differences in below ground biomass allocation. *Biomass and Bioenergy* 80, 114–27. doi:10.1016/j.biombioe.2015.04.020.

Dalin, P., Bjorkman, C. and Eklund, K. 2004. Leaf beetle grazing does not induce willow trichome defence in the coppicing willow *Salix viminalis*. *Agricultural and Forest Entomology* 6(2), 105–9. doi:10.1111/j.1461-9555.2004.00211.x.

Decker, S. R., Brunecky, R., Tucker, M. P., Himmel, M. E. and Selig, M. J. 2009. High-throughput screening techniques for biomass conversion. *Bioenergy Research* 2(4), 179–92. doi:10.1007/s12155-009-9051-0.

Desiree, J. I., Pita, A. V., Floor, V. D. H. and Andre, P. C. F. 2014. Biodiversity impacts of bioenergy crop production: a state-of-the-art review. *Global Change Biology Bioenergy* 6, 183–209.

Dickmann, D. I. and Kuzovkina, J. 2014. Poplars and willows of the world, with emphasis on silviculturally important species. In: Isebrands, J. G. and Richardson, J. (Eds), *Poplars and Willows: Trees for Society and the Environment*. CABI Publishing, Wallingford.

Dillen, M., Vanhellemont, M., Verdonckt, P., Maes, W. H., Steppe, K. and Verheyen, K. 2016. Productivity, stand dynamics and the selection effect in a mixed willow clone short rotation coppice plantation. *Biomass and Bioenergy* 87, 46–54. doi:10.1016/j.biombioe.2016.02.013.

Dimitriou, I. and Aronsson, P. 2005. Willows for energy and phytoremediation in Sweden. *Unasylva (English Education)* 56, 47–50.

Dimitriou, I. and Mola-Yudego, B. 2017. Poplar and willow plantations on agricultural land in Sweden: area, yield, groundwater quality and soil organic carbon. *Forest Ecology and Management* 383, 99–107. doi:10.1016/j.foreco.2016.08.022.

Dimitriou, I., Baum, C., Baum, S., Busch, G., Schulz, U., Kohn, J., Lamersdorf, N., Leinweber, P., Aronsson, P., Weih, M., et al. 2009. The impact of Short Rotation Coppice (SRC) cultivation on the environment. *Landbauforschung Volkenrode* 59, 159-62.

Dimitriou, I., Mola-Yudego, B. and Aronsson, P. 2012a. Impact of willow short rotation coppice on water quality. *Bioenergy Research* 5(3), 537–45. doi:10.1007/s12155-012-9211-5.

Dimitriou, I., Mola-Yudego, B., Aronsson, P. and Eriksson, J. 2012b. Changes in organic carbon and trace elements in the soil of willow short-rotation coppice plantations. *Bioenergy Research* 5(3), 563–72. doi:10.1007/s12155-012-9215-1.

Edelfeldt, S., Lundkvist, A., Forkman, J. and Verwijst, T. 2015. Effects of cutting length, orientation and planting depth on early willow shoot establishment. *Bioenergy Research* 8(2), 796–806. doi:10.1007/s12155-014-9560-3.

Ericsson, N., Porso, C., Ahlgren, S., Nordberg, Å, Sundberg, C. and Hansson, P. A. 2013. Time-dependent climate impact of a bioenergy system – methodology development and application to Swedish conditions. *Global Change Biology Bioenergy* 5(5), 580–90. doi:10.1111/gcbb.12031.

Ericsson, N., Sundberg, C., Nordberg, Å, Ahlgren, S. and Hansson, P. A. 2017. Time-dependent climate impact and energy efficiency of combined heat and power production from short-rotation coppice willow using pyrolysis or direct combustion. *Global Change Biology Bioenergy* 9(5), 876–90. doi:10.1111/gcbb.12415.

Estevez, M. M., Linjordet, R. and Morken, J. 2012. Effects of steam explosion and co-digestion in the methane production from *Salix* by mesophilic batch assays. *Bioresource Technology* 104, 749–56. doi:10.1016/j.biortech.2011.11.017.

Estevez, M. M., Sapci, Z., Linjordet, R., Schnurer, A. and Morken, J. 2014. Semi-continuous anaerobic co-digestion of cow manure and steam-exploded *Salix* with recirculation of liquid digestate. *Journal of Environmental Management* 136, 9–15. doi:10.1016/j.jenvman.2014.01.028.

Fabio, E. S. and Smart, L. B. 2018a. Differential growth response to fertilization of ten elite shrub willow (*Salix* spp.) bioenergy cultivars. *Trees-Structure and Function* 32(4), 1061–72. doi:10.1007/s00468-018-1695-y.

Fabio, E. S. and Smart, L. B. 2018b. Effects of nitrogen fertilization in shrub willow short rotation coppice production – a quantitative review. *Global Change Biology Bioenergy* 10(8), 548–64. doi:10.1111/gcbb.12507.

Fabio, E. S., Volk, T. A., Miller, R. O., Serapiglia, M. J., Kemanian, A. R., Montes, F., Kuzovkina, Y. A., Kling, G. J. and Smart, L. B. 2017. Contributions of environment and genotype to variation in shrub willow biomass composition. *Industrial Crops and Products* 108, 149–61. doi:10.1016/j.indcrop.2017.06.030.

FAO. 2016. *Poplars and Other Fast-Growing Trees – Renewable Resources for Future Green Economies. Synthesis of Country Progress Reports.* Forestry Policy and Resources Division, Food and Agriculture Organization, Rome, Italy.

Fransson, P. M. A., Toljander, Y. K., Baum, C. and Weih, M. 2013. Host plant–ectomycorrhizal fungus combination drives resource allocation in willow: evidence for complex species interaction from a simple experiment. *Ecoscience* 20(2), 112-21. doi:10.2980/20-2-3576.

Ghelardini, L., Berlin, S., Weih, M., Lagercrantz, U., Gyllenstrand, N. and Ronnberg-Wastljung, A. C. Y. 2014. Genetic architecture of spring and autumn phenology in Salix. *BMC Plant Biology* 14, 31. doi:10.1186/1471-2229-14-31.

Glynn, C., Ronnberg-Wastljung, A., Julkunen-Tiitto, R. and Weih, M. 2004. Willow genotype, but not drought treatment, affects foliar phenolic concentrations and leaf-beetle resistance. *Entomologia Experimentalis et Applicata* 113(1), 1–14. doi:10.1111/j.0013-8703.2004.00199.x.

Glynn, C., Herms, D. A., Orians, C. M., Hansen, R. C. and Larsson, S. 2007. Testing the growth-differentiation balance hypothesis: dynamic responses of willows to nutrient availability. *The New Phytologist* 176(3), 623–34. doi:10.1111/j.1469-8137.2007.02203.x.

Gonzalez-Garcia, S., Mola-Yudego, B., Dimitriou, I., Aronsson, P. and Murphy, R. 2012. Environmental assessment of energy production based on long term commercial willow plantations in Sweden. *The Science of the Total Environment* 421–422, 210-9. doi:10.1016/j.scitotenv.2012.01.041.

Gregory, A. S., Dungait, J. A. J., Shield, I. F., Macalpine, W. J., Cunniff, J., Durenkamp, M., White, R. P., Joynes, A. and Richter, G. M. 2018. Species and genotype effects of bioenergy crops on root production, carbon and nitrogen in temperate agricultural soil. *Bioenergy Research* 11(2), 382–97. doi:10.1007/s12155-018-9903-6.

Grogan, P. and Matthews, R. 2002. A modelling analysis of the potential for soil carbon sequestration under short rotation coppice willow bioenergy plantations. *Soil Use and Management* 18(3), 175–83. doi:10.1111/j.1475-2743.2002.tb00237.x.

Grossman, J. J., Vanhellemont, M., Barsoum, N., Bauhus, J., Bruelheide, H., Castagneyrol, B., Cavender-Bares, J., Eisenhauer, N., Ferlian, O., Gravel, D., et al. 2018. Synthesis and future research directions linking tree diversity to growth, survival, and damage in a global network of tree diversity experiments. *Environmental and Experimental Botany* 152, 68–89. doi:10.1016/j.envexpbot.2017.12.015.

Gullberg, U. 1993. Towards making willows pilot species for coppicing production. *The Forestry Chronicle* 69(6), 721-6. doi:10.5558/tfc69721-6.

Hallingback, H. R., Fogelqvist, J., Powers, S. J., Turrion-Gomez, J., Rossiter, R., Amey, J., Martin, T., Weih, M., Gyllenstrand, N., Karp, A., et al. 2016. Association mapping in *Salix viminalis* L. (Salicaceae) – identification of candidate genes associated with growth and phenology. *Global Change Biology. Bioenergy* 8(3), 670–85. doi:10.1111/gcbb.12280.

Hammar, T., Ericsson, N., Sundberg, C. and Hansson, P. A. 2014. Climate impact of willow grown for bioenergy in Sweden. *Bioenergy Research* 7(4), 1529–40. doi:10.1007/s12155-014-9490-0.

Hammar, T., Hansson, P. A. and Sundberg, C. 2017. Climate impact assessment of willow energy from a landscape perspective: a Swedish case study. *Global Change Biology Bioenergy* 9(5), 973–85. doi:10.1111/gcbb.12399.

Hanley, S. J. and Karp, A. 2014. Genetic strategies for dissecting complex traits in biomass willows (*Salix* spp.). *Tree Physiology* 34(11), 1167–80. doi:10.1093/treephys/tpt089.

Hanley, S. J., Mallott, M. D. and Karp, A. 2006. Alignment of a *Salix* linkage map to the *Populus* genomic sequence reveals macrosynteny between willow and poplar genomes. *Tree Genetics and Genomes* 3(1), 35–48. doi:10.1007/s11295-006-0049-x.

Harris, Z. M., Spake, R. and Taylor, G. 2015. Land use change to bioenergy: a meta-analysis of soil carbon and GHG emissions. *Biomass and Bioenergy* 82, 27–39. doi:10.1016/j.biombioe.2015.05.008.

Harris, Z. M., Alberti, G., Viger, M., Jenkins, J. R., Rowe, R., Mcnamara, N. P. and Taylor, G. 2017. Land-use change to bioenergy: grassland to short rotation coppice willow has an improved carbon balance. *Global Change Biology Bioenergy* 9(2), 469–84. doi:10.1111/gcbb.12347.

Hartley, M. J. 2002. Rationale and methods for conserving biodiversity in plantation forests. *Forest Ecology and Management* 155(1-3), 81–95. doi:10.1016/S0378-1127(01)00549-7.

Heller, M. C., Keoleian, G. A. and Volk, T. A. 2003. Life cycle assessment of a willow bioenergy cropping system. *Biomass and Bioenergy* 25(2), 147–65. doi:10.1016/S0961-9534(02)00190-3.

Higman, S., Mayers, J., Bass, S., Judd, N. and Nussbaum, R. 2005. *The Sustainable Forestry Handbook*. Earthscan Publications, London, UK.

Hoeber, S., Fransson, P., Prieto-Ruiz, I., Manzoni, S. and Weih, M. 2017. Two *Salix* genotypes differ in productivity and nitrogen economy when grown in monoculture and mixture. *Frontiers in Plant Science* 8, 231. doi:10.3389/fpls.2017.00231.

Hoeber, S., Arranz, C., Nordh, N. E., Baum, C., Low, M., Nock, C., Scherer-Lorenzen, M. and Weih, M. 2018. Genotype identity has a more important influence than genotype diversity on shoot biomass productivity in willow short-rotation coppices. *Global Change Biology Bioenergy* 10(8), 534–47. doi:10.1111/gcbb.12521.

Hoglund, S. and Larsson, S. 2005. Abiotic induction of susceptibility in insect-resistant willow. *Entomologia Experimentalis et Applicata* 115(1), 89–96. doi:10.1111/j.1570-7458.2005.00260.x.

Hoglund, S., Larsson, S. and Wingsle, G. 2005. Both hypersensitive and non-hypersensitive responses are associated with resistance in *Salix viminalis* against the gall midge

*Dasineura marginemtorquens*. *Journal of Experimental Botany* 56(422), 3215–22. doi:10.1093/jxb/eri318.

Holden, E., Linnerud, K. and Banister, D. 2014. Sustainable development: our common future revisited. *Global Environmental Change-Human and Policy Dimensions* 26, 130–9. doi:10.1016/j.gloenvcha.2014.04.006.

Horn, S. J., Estevez, M. M., Nielsen, H. K., Linjordet, R. and Eijsink, V. G. H. 2011. Biogas production and saccharification of *Salix* pretreated at different steam explosion conditions. *Bioresource Technology* 102(17), 7932–6. doi:10.1016/j.biortech.2011.06.042.

Hrynkiewicz, K., Baum, C., Leinweber, P., Weih, M. and Dimitriou, I. 2010. The significance of rotation periods for mycorrhiza formation in Short Rotation Coppice. *Forest Ecology and Management* 260(11), 1943–9. doi:10.1016/j.foreco.2010.08.020.

Hrynkiewicz, K., Toljander, Y. K., Baum, C., Fransson, P. M. A., Taylor, A. F. S. and Weih, M. 2012. Correspondence of ectomycorrhizal diversity and colonisation of willows (*Salix* spp.) grown in short rotation coppice on arable sites and adjacent natural stands. *Mycorrhiza* 22(8), 603–13. doi:10.1007/s00572-012-0437-z.

Isebrands, J. G., Aronsson, P., Carlson, M., Ceulemans, R., Coleman, M., Dickinson, N., Dimitriou, J., Doty, S., Gardiner, E., Heinsoo, K., et al. 2014. Environmental applications of poplars and willows. In: Isebrands, J. G. and Richardson, J. (Eds), *Poplars and Willows: Trees for Society and the Environment*. CABI Publishing, Wallingford.

Jaworska, J. 2010. The biodiversity of flying beneficial insects on a short rotation willow. *Progress in Plant Protection* 50, 1486–90.

Jirjis, R. 2005. Effects of particle size and pile height on storage and fuel quality of comminuted *Salix viminalis*. *Biomass and Bioenergy* 28(2), 193–201. doi:10.1016/j.biombioe.2004.08.014.

Jones, M. D., Durall, D. M. and Tinker, P. B. 1991. Fluxes of carbon and phosphorus between symbionts in willow ectomycorrhizas and their changes with time. *New Phytologist* 119(1), 99–106. doi:10.1111/j.1469-8137.1991.tb01012.x.

Kahle, P., Baum, C. and Boelcke, B. 2005. Effect of afforestation on soil properties and mycorrhizal formation. *Pedosphere* 15, 754–60.

Karp, A. and Shield, I. 2008. Bioenergy from plants and the sustainable yield challenge. *The New Phytologist* 179(1), 15–32. doi:10.1111/j.1469-8137.2008.02432.x.

Karp, A., Hanley, S. J., Trybush, S. O., Macalpine, W., Pei, M. and Shield, I. 2011. Genetic improvement of willow for bioenergy and biofuels. *Journal of Integrative Plant Biology* 53(2), 151–65. doi:10.1111/j.1744-7909.2010.01015.x.

Krzyzaniak, M., Stolarski, M. J., Waliszewska, B., Szczukowski, S., Tworkowski, J., Zaluski, D. and Snieg, M. 2014. Willow biomass as feedstock for an integrated multi-product biorefinery. *Industrial Crops and Products* 58, 230–7. doi:10.1016/j.indcrop.2014.04.033.

Krzyzaniak, M., Stolarski, M. J., Niksa, D., Tworkowski, J. and Szczukowski, S. 2016a. Effect of storage methods on willow chips quality. *Biomass and Bioenergy* 92, 61–9. doi:10.1016/j.biombioe.2016.06.007.

Krzyzaniak, M., Stolarski, M. J., Szczukowski, S. and Tworkowski, J. 2016b. Life cycle assessment of new willow cultivars grown as feedstock for integrated biorefineries. *Bioenergy Research* 9(1), 224–38. doi:10.1007/s12155-015-9681-3.

Kuzovkina, Y. A., Weih, M., Abalos Romero, M., Charles, J., Hurst, S., Mcivor, I., Karp, A., Trybush, S., Labrecque, M. and Teodorescu, T. I. 2008. *Salix*: botany and global horticulture. *Horticultural Reviews* 34, 447–89.

Labrecque, M. and Teodorescu, T. I. 2003. High biomass yield achieved by *Salix* clones in SRIC following two 3-year coppice rotations on abandoned farmland in southern Quebec, Canada. *Biomass and Bioenergy* 25(2), 135–46. doi:10.1016/S0961-9534(02)00192-7.

Langeveld, H., Quist-Wessel, F., Dimitriou, I., Aronsson, P., Baum, C., Schulz, U., Bolte, A., Baum, S., Kohn, J., Weih, M., et al. 2012. Assessing environmental impacts of short rotation coppice (SRC) expansion: model definition and preliminary results. *Bioenergy Research* 5(3), 621–35. doi:10.1007/s12155-012-9235-x.

Larsson, S. 1997. Commercial breeding of willow for short rotation coppice. *Aspects of Applied Biology* 49, 215–8.

Larsson, S., Nordh, N. E., Farrell, J. and Tweddle, P. 2007. Manual for SRC willow growers. Lantmännen Agroenergi AB, Örebro, Sweden. Available at: www.agroenergi.se.

Ledin, S. 1996. Willow wood properties, production and economy. *Biomass and Bioenergy* 11(2-3), 75–83. doi:10.1016/0961-9534(96)00022-0.

Lennartsson, M. and Ogren, E. 2004. Clonal variation in temperature requirements for budburst and dehardening in *Salix* species used for biomass production. *Scandinavian Journal of Forest Research* 19(4), 295–302. doi:10.1080/02827580410030145.

Liang, J. J., Crowther, T. W., Picard, N., Wiser, S., Zhou, M., Alberti, G., Schulze, E. D., McGuire, A. D., Bozzato, F., Pretzsch, H., et al. 2016. Positive biodiversity-productivity relationship predominant in global forests. *Science* 354(6309), 12. doi:10.1126/science.aaf8957.

Lindroth, A. and Bath, A. 1999. Assessment of regional willow coppice yield in Sweden on basis of water availability. *Forest Ecology and Management* 121(1-2), 57–65. doi:10.1016/S0378-1127(98)00556-8.

Marron, N., Gielen, B., Brignolas, F., Gao, J., Johnson, J. D., Karnosky, D. F., Polle, A., Scarascia-Mugnozza, G., Schroeder, W. R. and Ceulemans, R. 2014. Abiotic stresses. In: Isebrands, J. G. and Richardson, J. (Eds), *Poplars and Willows: Trees for Society and the Environment*. CABI Publishing, Wallingford.

McCracken, A. R. and Dawson, W. M. 1994. Experiences in the use of mixed-clonal stands of *Salix* as a method of reducing the impact of rust diseases. *Norwegian Journal of Agricultural Sciences (Suppl.)* (18), 101–9.

McCracken, A. R. and Dawson, W. M. 1998. Interaction of willow (*Salix*) clones growing in mixtures. *Annals of Applied Biology* 132, 54–5.

Mitchell, C. P., Ford-Robertson, J. B., Hinckley, T. and Sennerby, F. L. (Eds). 1992. *Ecophysiology of Short Rotation Forest Crops*. Elsevier Science Publishers Ltd, Barking, UK.

Mola-Yudego, B. 2011. Trends and productivity improvements from commercial willow plantations in Sweden during the period 1986–2000. *Biomass and Bioenergy* 35(1), 446–53. doi:10.1016/j.biombioe.2010.09.004.

Mola-Yudego, B. and Aronsson, P. 2008. Yield models for commercial willow biomass plantations in Sweden. *Biomass and Bioenergy* 32(9), 829–37. doi:10.1016/j.biombioe.2008.01.002.

Moritz, K. K., Bjorkman, C., Parachnowitsch, A. L. and Stenberg, J. A. 2017. Plant sex effects on insect herbivores and biological control in a Short Rotation Coppice willow. *Biological Control* 115, 30–6. doi:10.1016/j.biocontrol.2017.09.006.

Muller, M., Klein, A. M., Scherer-Lorenzen, M., Nock, C. A. and Staab, M. 2018. Tree genetic diversity increases arthropod diversity in willow short rotation coppice. *Biomass and Bioenergy* 108, 338–44. doi:10.1016/j.biombioe.2017.12.001.

Nordborg, M., Berndes, G., Dimitriou, I., Henriksson, A., Mola-Yudego, B. and Rosenqvist, H. 2018. Energy analysis of willow production for bioenergy in Sweden. *Renewable and Sustainable Energy Reviews* 93, 473–82. doi:10.1016/j.rser.2018.05.045.

Nordh, N. E. 2005. Long term changes in stand structure and biomass production in short rotation willow coppice. Doctoral thesis. Swedish University of Agricultural Sciences.

Ogren, E. 1999. Fall frost resistance in willows used for biomass production. II. Predictive relationships with sugar concentration and dry matter content. *Tree Physiology* 19(11), 755–60. doi:10.1093/treephys/19.11.755.

Ostry, M., Ramstedt, M., Newcombe, G. and Steenackers, M. 2014. Diseases of poplars and willows. In: Isebrands, J. G. and Richardson, J. (Eds), *Poplars and Willows: Trees for Society and the Environment*. CABI Publishing, Wallingford.

Parajuli, R., Knudsen, M. T., Djomo, S. N., Corona, A., Birkved, M. and Dalgaard, T. 2017. Environmental life cycle assessment of producing willow, alfalfa and straw from spring barley as feedstocks for bioenergy or biorefinery systems. *The Science of the Total Environment* 586, 226–40. doi:10.1016/j.scitotenv.2017.01.207.

Pawar, P. M. A., Schnurer, A., Mellerowicz, E. J. and Ronnberg-Wastljung, A. C. 2018. QTL mapping of wood FT-IR chemotypes shows promise for improving biofuel potential in short rotation coppice willow (*Salix* spp.). *Bioenergy Research* 11(2), 351-63. doi:10.1007/s12155-018-9901-8.

Pei, M. H., Lindegaard, K., Ruiz, C. and Bayon, C. 2008. Rust resistance of some varieties and recently bred genotypes of biomass willows. *Biomass and Bioenergy* 32(5), 453–9. doi:10.1016/j.biombioe.2007.11.002.

Philippot, S. 1996. Simulation models of short-rotation forestry production and coppice biology. *Biomass and Bioenergy* 11(2–3), 85–93. doi:10.1016/0961-9534(96)00008-6.

Phitsuwan, P., Sakka, K. and Ratanakhanokchai, K. 2013. Improvement of lignocellulosic biomass in planta: a review of feedstocks, biomass recalcitrance, and strategic manipulation of ideal plants designed for ethanol production and processability. *Biomass and Bioenergy* 58, 390–405. doi:10.1016/j.biombioe.2013.08.027.

Pucholt, P., Sjodin, P., Weih, M., Ronnberg-Wastljung, A. C. and Berlin, S. 2015. Genome-wide transcriptional and physiological responses to drought stress in leaves and roots of two willow genotypes. *BMC Plant Biology* 15, 244-. doi:10.1186/s12870-015-0630-2.

Puentes, A., Torp, M., Weih, M. and Bjorkman, C. 2015. Direct effects of elevated temperature on a tri-trophic system: *Salix*, leaf beetles and predatory bugs. *Arthropod-Plant Interactions* 9(6), 567–75. doi:10.1007/s11829-015-9401-0.

Ra, K., Shiotsu, F., Abe, J. and Morita, S. 2012. Biomass yield and nitrogen use efficiency of cellulosic energy crops for ethanol production. *Biomass and Bioenergy* 37, 330–4. doi:10.1016/j.biombioe.2011.12.047.

Ray, M. J., Brereton, N. J. B., Shield, I., Karp, A. and Murphy, R. J. 2012. Variation in cell wall composition and accessibility in relation to biofuel potential of short rotation coppice willows. *Bioenergy Research* 5(3), 685–98. doi:10.1007/s12155-011-9177-8.

Richardson, J., Isebrands, J. G. and Ball, J. B. 2014. Ecology and physiology of poplars and willows. In: Isebrands, J. G. and Richardson, J. (Eds), *Poplars and Willows: Trees for Society and the Environment*. CABI Publishing, Wallingford.

Ronnberg-Wastljung, A. C. 2001. Genetic structure of growth and phenological traits in *Salix viminalis*. *Canadian Journal of Forest Research* 31(2), 276–82. doi:10.1139/x00-175.

Ronnberg-Wastljung, A. C. and Gullberg, U. 1999. Genetics of breeding characters with possible effects on biomass production in *Salix viminalis* (L.). *Theoretical and Applied Genetics* 98(3-4), 531–40. doi:10.1007/s001220051101.

Ronnberg-Wastljung, A. C., Glynn, C. and Weih, M. 2005. QTL analyses of drought tolerance and growth for a *Salix dasyclados* × *Salix viminalis* hybrid in contrasting water regimes. *Theoretical and Applied Genetics* 110(3), 537–49. doi:10.1007/s00122-004-1866-7.

Rooney, D. C., Killham, K., Bending, G. D., Baggs, E., Weih, M. and Hodge, A. 2009. Mycorrhizas and biomass crops: opportunities for future sustainable development. *Trends in Plant Science* 14(10), 542-9. doi:10.1016/j.tplants.2009.08.004.

Samils, B., Ronnberg-Wastljung, A. C. and Stenlid, J. 2011. QTL mapping of resistance to leaf rust in *Salix*. *Tree Genetics and Genomes* 7(6), 1219-35. doi:10.1007/s11295-011-0408-0.

Sandak, J. and Sandak, A. 2011. Fourier transform near infrared assessment of biomass composition of shrub willow clones (*Salix* sp.) for optimal bio-conversion processing. *Journal of Near Infrared Spectroscopy* 19(5), 309–18. doi:10.1255/jnirs.950.

Sandak, A., Sandak, J., Waliszewska, B., Zborowska, M. and Mleczek, M. 2017. Selection of optimal conversion path for willow biomass assisted by near infrared spectroscopy. *iForest - Biogeosciences and Forestry* 10(2), 506–14. doi:10.3832/ifor1987-010.

Sannervik, A. N., Eckersten, H., Verwijst, T., Kowalik, P. and Nordh, N. E. 2006. Simulation of willow productivity based on radiation use efficiency, shoot mortality and shoot age. *European Journal of Agronomy* 24(2), 156–64. doi:10.1016/j.eja.2005.07.007.

Sassner, P., Galbe, M. and Zacchi, G. 2008a. Techno-economic evaluation of bioethanol production from three different lignocellulosic materials. *Biomass and Bioenergy* 32(5), 422–30. doi:10.1016/j.biombioe.2007.10.014.

Sassner, P., Martensson, C. G., Galbe, M. and Zacchi, G. 2008b. Steam pretreatment of H2SO4-impregnated *Salix* for the production of bioethanol. *Bioresource Technology* 99(1), 137–45. doi:10.1016/j.biortech.2006.11.039.

Schmidt-Walter, P. and Lamersdorf, N. P. 2012. Biomass production with willow and poplar short rotation coppices on sensitive areas-the impact on nitrate leaching and groundwater recharge in a drinking water catchment near Hanover, Germany. *Bioenergy Research* 5(3), 546–62. doi:10.1007/s12155-012-9237-8.

Schulz, U., Brauner, O. and Gruss, H. 2009. Animal diversity on short-rotation coppices - a review. *Landbauforschung Volkenrode* 59, 171–81.

Serapiglia, M. J., Cameron, K. D., Stipanovic, A. J. and Smart, L. B. 2009. Analysis of biomass composition using high-resolution thermogravimetric analysis and percent bark content for the selection of shrub willow bioenergy crop varieties. *Bioenergy Research* 2(1-2), 1–9. doi:10.1007/s12155-008-9028-4.

Serapiglia, M. J., Cameron, K. D., Stipanovic, A. J. and Smart, L. B. 2012. Correlations of expression of cell wall biosynthesis genes with variation in biomass composition in shrub willow (*Salix* spp.) biomass crops. *Tree Genetics and Genomes* 8(4), 775–88. doi:10.1007/s11295-011-0462-7.

Serapiglia, M. J., Cameron, K. D., Stipanovic, A. J., Abrahamson, L. P., Volk, T. A. and Smart, L. B. 2013a. Yield and woody biomass traits of novel shrub willow hybrids at two contrasting sites. *Bioenergy Research* 6(2), 533–46. doi:10.1007/s12155-012-9272-5.

Serapiglia, M. J., Humiston, M. C., Xu, H. W., Hogsett, D. A., De Orduna, R. M., Stipanovic, A. J. and Smart, L. B. 2013b. Enzymatic saccharification of shrub willow genotypes with differing biomass composition for biofuel production. *Frontiers in Plant Science* 4, 57. doi:10.3389/fpls.2013.00057.

Serapiglia, M. J., Gouker, F. E., Hart, J. F., Unda, F., Mansfield, S. D., Stipanovic, A. J. and Smart, L. B. 2015. Ploidy level affects important biomass traits of novel shrub Willow (*Salix*) hybrids. *Bioenergy Research* 8(1), 259–69. doi:10.1007/s12155-014-9521-x.

Smart, L. B. and Cameron, K. D. 2008. Genetic improvement of willow (*Salix* spp.) as a dedicated bioenergy crop. In: Vermerris, W. (Ed.), *Genetic Improvement of Bioenergy Crops*. Springer, New York.

Stanton, B. J., Serapiglia, M. J. and Smart, L. B. 2014. The domestication and conservation of *Populus* and *Salix* genetic resources. In: Isebrands, J. G. and Richardson, J. (Eds), *Poplars and Willows: Trees for Society and the Environment*. CABI Publishing, Wallingford.

Stanturf, J. A. and Van Oosten, C. 2014. Operational poplar and willow culture. In: Isebrands, J. G. and Richardson, J. (Eds), *Poplars and Willows: Trees for Society and the Environment*. CABI Publishing, Wallingford.

Stenberg, J. A., Lehrman, A. and Bjorkman, C. 2010. Uncoupling direct and indirect plant defences: novel opportunities for improving crop security in willow plantations. *Agriculture, Ecosystems and Environment* 139(4), 528–33. doi:10.1016/j.agee.2010.09.013.

Stolarski, M. J., Szczukowski, S., Tworkowski, J., Wroblewska, H. and Krzyzaniak, M. 2011. Short rotation willow coppice biomass as an industrial and energy feedstock. *Industrial Crops and Products* 33(1), 217–23. doi:10.1016/j.indcrop.2010.10.013.

Stolarski, M. J., Szczukowski, S., Tworkowski, J. and Klasa, A. 2013. Yield, energy parameters and chemical composition of short-rotation willow biomass. *Industrial Crops and Products* 46, 60–5. doi:10.1016/j.indcrop.2013.01.012.

Szczukowski, S., Tworkowski, J., Klasa, A. and Stolarski, M. 2002. Productivity and chemical composition of wood tissues of short rotation willow coppice cultivated on arable land. *Rostlinna Vyroba* 48, 413–7.

Tilman, D., Cassman, K. G., Matson, P. A., Naylor, R. and Polasky, S. 2002. Agricultural sustainability and intensive production practices. *Nature* 418(6898), 671–7. doi:10.1038/nature01014.

Tsarouhas, V., Gullberg, U. and Lagercrantz, U. 2004. Mapping of quantitative trait loci (QTLs) affecting autumn freezing resistance and phenology in *Salix*. *Theoretical and Applied Genetics* 108(7), 1335–42. doi:10.1007/s00122-003-1544-1.

Volk, T. A., Verwijst, T., Tharakan, P. J., Abrahamson, L. P. and White, E. H. 2004. Growing fuel: a sustainability assessment of willow biomass crops. *Frontiers in Ecology and the Environment* 2(8), 411–8. doi:10.1890/1540-9295(2004)002[0411:GFASAO]2.0.CO;2.

Volk, T. A., Heavey, J. P. and Eisenbies, M. H. 2016. Advances in shrub-willow crops for bioenergy, renewable products, and environmental benefits. *Food and Energy Security* 5(2), 97–106. doi:10.1002/fes3.82.

Wang, D., Jaiswal, D., Lebauer, D. S., Wertin, T. M., Bollero, G. A., Leakey, A. D. B. and Long, S. P. 2015. A physiological and biophysical model of coppice willow (*Salix* spp.) production yields for the contiguous USA in current and future climate scenarios. *Plant, Cell and Environment* 38(9), 1850–65. doi:10.1111/pce.12556.

WCED. 1987. *Our Common Future*. Oxford University Press, Oxford.

Weih, M. 2004. Intensive short rotation forestry in boreal climates: present and future perspectives. *Canadian Journal of Forest Research-Revue Canadienne De Recherche Forestiere* 34(7), 1369–78. doi:10.1139/x04-090.

Weih, M. 2009. Genetic and environmental variation in spring and autumn phenology of biomass willows (*Salix* spp.): effects on shoot growth and nitrogen economy. *Tree Physiology* 29(12), 1479-90. doi:10.1093/treephys/tpp081.

Weih, M. 2013. Willow. In: Singh, B. P. (Ed.), *Biofuel Crops: Production, Physiology and Genetics*. CABI Publishing, Wallingford, UK.

Weih, M. and Dimitriou, I. 2012. Environmental impacts of short rotation coppice (SRC) grown for biomass on agricultural land. *BioEnergy Research* 5(3), 535-6. doi:10.1007/s12155-012-9230-2.

Weih, M. and Nordh, N. E. 2002. Characterising willows for biomass and phytoremediation: growth, nitrogen and water use of 14 willow clones under different irrigation and fertilisation regimes. *Biomass and Bioenergy* 23(6), 397-413. doi:10.1016/S0961-9534(02)00067-3.

Weih, M. and Nordh, N. E. 2005. Determinants of biomass production in hybrid willows and prediction of field performance from pot studies. *Tree Physiology* 25(9), 1197-206. doi:10.1093/treephys/25.9.1197.

Weih, M. and Nordh, N. E. 2009. *Biomass Production With Fast-Growing Trees on Agricultural Land in Cool-Temperate Regions: Possibilities, Limitations, Challenges*. Nova Science Publishers, Inc, Hauppauge.

Weih, M. and Ronnberg-Wastljung, A. C. 2007. Shoot biomass growth is related to the vertical leaf nitrogen gradient in Salix canopies. *Tree Physiology* 27(11), 1551-9. doi:10.1093/treephys/27.11.1551.

Weih, M., Ronnberg-Wastljung, A. C. and Glynn, C. 2006. Genetic basis of phenotypic correlations among growth traits in hybrid willow (*Salix dasyclados* × *S-viminalis*) grown under two water regimes. *The New Phytologist* 170(3), 467-77. doi:10.1111/j.1469-8137.2006.01685.x.

Weih, M., Didon, U. M. E., Ronnberg-Wastljung, A.-C. and Bjorkman, C. 2008. Integrated agricultural research and crop breeding: allelopathic weed control in cereals and long-term productivity in perennial bion ass crops. *Agricultural Systems* 97(3), 99-107. doi:10.1016/j.agsy.2008.02.009.

Weih, M., Bonosi, L., Ghelardini, L. and Ronnberg-Wastljung, A. C. 2011. Optimizing nitrogen economy under drought: increased leaf nitrogen is an acclimation to water stress in willow (*Salix* spp.). *Annals of Botany* 108(7), 1347-53. doi:10.1093/aob/mcr227.

Weih, M., Hoeber, S., Beyer, F. and Fransson, P. 2014. Traits to ecosystems: the ecological sustainability challenge when developing future energy crops. *Frontiers in Energy Research* 2, 1-5. doi:10.3389/fenrg.2014.00017.

Weih, M., Hamner, K. and Pourazari, F. 2018. Analyzing plant nutrient uptake and utilization efficiencies: comparison between crops and approaches. *Plant and Soil* 430(1-2), 7-21. doi:10.1007/s11104-018-3738-y.

Welc, M., Lundkvist, A. and Verwijst, T. 2017. Effects of cutting phenology (non-dormant versus dormant) on early growth performance of three willow clones grown under different weed treatments and planting dates. *Bioenergy Research* 10(4), 1094-104. doi:10.1007/s12155-017-9871-2.

Verheyen, K., Vanhellemont, M., Auge, H., Baeten, L., Baraloto, C., Barsoum, N., Bilodeau-Gauthier, S., Bruelheide, H., Castagneyrol, B., Godbold, D., et al. 2016. Contributions of a global network of tree diversity experiments to sustainable forest plantations. *Ambio* 45(1), 29-41. doi:10.1007/s13280-015-0685-1.

Verwijst, T., Lundkvist, A., Edelfeldt, S. and Albertsson, J. 2013. Development of sustainable willow short rotation forestry in northern Europe. In: Matovic, M. D. (Ed.), *Biomass Now*. IntechOpen. doi:10.5772/55072. Available at: https://www.intechopen.com/books/biomass-now-sustainable-growth-and-use/development-of-sustainable-willow-short-rotation-forestry-in-northern-europe.

Whittaker, C., Yates, N. E., Powers, S. J., Misselbrook, T. and Shield, I. 2018. Dry matter losses and quality changes during short rotation coppice willow storage in chip or rod form. *Biomass and Bioenergy* 112, 29–36. doi:10.1016/j.biombioe.2018.02.005.

Xue, C., Penton, C. R., Zhang, B., Zhao, M., Rothstein, D. E., Mladenoff, D. J., Forrester, J. A., Shen, Q. and Tiedje, J. M. 2016. Soil fungal and bacterial responses to conversion of open land to short-rotation woody biomass crops. *Global Change Biology Bioenergy* 8(4), 723–36. doi:10.1111/gcbb.12303.

# Chapter 5

## The importance of agroforestry systems in supporting biodiversity conservation and agricultural production: a European perspective

*M. R. Mosquera-Losada, J. J. Santiago-Freijanes, A. Rigueiro-Rodríguez, F. J. Rodríguez-Rigueiro, D. Arias Martínez, A. Pantera and N. Ferreiro-Domínguez, University of Santiago de Compostela, Spain*



## 1 Introduction

Biodiversity has become a key international environmental and production issue in recent decades, as shown by the Convention on Biological Diversity (CBD) agreed at the Rio de Janeiro Summit in 1992 (UN 1992). More recently, action on biodiversity has been reinforced with a global strategic plan to combat biodiversity losses (Aichi targets (CBD 2010)) as well as other initiatives such as the EU's 2020 biodiversity strategy (EC 2011b).

Biodiversity is part of the natural capital that needs to be preserved for the following generations. Millions of years of evolution are represented in many different types of organisms that may be useful for both food and health. As an example, most traditional medicines are based on the extraction of existing plant compounds. This means that most pharmaceutical compounds are not

http://dx.doi.org/10.19103/AS.2020.0071.14

'created' by the pharmaceutical industry. Instead they are 'copied' from nature which means we need the 'originals' to be preserved.

Biodiversity is crucial to sustainably increase crop production per unit of land by optimal combination of plants that are best able to use soil and other natural resources in a complementary way and thus contribute to the provision of ecosystem services. The co-evolution of different species adapted to a specific habitat has been ignored by intensive agricultural practices and therefore the complementary capacity they have to use resources optimally and provide ecosystem services has also been ignored.

The reduction in biodiversity can be seen, for example, in Europe where only 17% of assessed habitat and 17% of assessed species have a favourable conservation status (Maes (2017). It has been estimated that this loss of biodiversity is causing large economic losses, for example, the loss of pollination is estimated as causing losses of €15 billion per year in the EU) (EC 2011b). Based on these facts and figures, the EU developed a biodiversity strategy based on six main targets (EC 2011a) that include the:

(1) development of a network of natural habitats;
(2) ecosystem services restoration based on the development of a green infrastructure;
(3) ensuring sustainability of agriculture and forestry activities;
(4) safeguarding and protecting EU fish stocks;
(5) controlling invasive species; and
(6) stepping up the EU's contribution to concerted global action to avert biodiversity loss.

The EU strategy is in line with two major commitments made by EU leaders in March 2010:

- halting the loss of biodiversity in the EU by 2020; and
- protecting, valuing and restoring EU biodiversity and ecosystem services by 2050.

As an integral part of the Europe 2020 Strategy, the biodiversity strategy contributes towards the EU's resource efficiency objectives by ensuring that Europe's natural capital is managed sustainably, as well as contributing to climate change mitigation and adaptation goals by improving the resilience of ecosystems and the services they provide.

Agroforestry can directly contribute to targets 1, 2, 3 and 6 of the EU biodiversity strategy. Agroforestry is a sustainable land use system where woody perennials (trees or shrubs) are combined with agricultural production (Mosquera-Losada et al. 2018a, b). Agroforestry can provide solutions to meet

European and global biodiversity strategies for reducing biodiversity losses and increasing biodiversity, while sustainably increasing the efficiency of agricultural systems through the use of the natural capital (Rigueiro-Rodríguez et al. 2010). This chapter describes the main reasons for agroforestry being able to enhance biodiversity and fulfil European and global biodiversity targets.

## 2 The contribution of agroforestry to global biodiversity goals

The main contribution of agroforestry to biodiversity is land heterogeneity due to the presence of woody perennials in agricultural systems or the understorey in forestry systems (Rigueiro-Rodríguez et al. 2009). Agroforestry is a type of land use that can be linked to all type of agricultural/forestry lands. The main types of agroforestry are (Mosquera-Losada et al. 2018a):

- silvopasture;
- silvoarable cultivation;
- riparian buffer strips;
- forest farming; and
- homegardens.

Silvopasture and silvoarable practices are implemented when farmers combine woody perennials with grazing animals and arable lands, respectively (Mosquera-Losada et al. 2017, 2018a). Woody perennials combined with agricultural production and adjacent to water courses can be categorized as riparian buffer strips. Forest farming is associated with understory agricultural production. When woody perennials usually associated with fruit trees are combined with agricultural production (i.e. vegetables) in urban/semiurban areas, we talk about 'homegardens' (Table 1).

Agroforestry improves biodiversity and therefore the potential provision of ecosystem services. The presence of isolated trees in an arable land clearly creates a spatial and temporal micro-environment with different water and nutrient pools that favours some varieties/species over the others, resulting in differences in species composition between the areas influenced by woody perennials and the open sites. Different understory varieties/species are differently adapted to different woody perennials-influenced areas. The heterogeneity caused by the presence of woody perennials affects not only plants but also other taxa, resulting in a cascade effect promoting the provision of ecosystem services.

As an example of heterogeneity, the level of adaptation of cereal crops to shade has been found to vary upon variety (Mosquera-Losada et al. 2019). The use of different woody perennials species in hedgerows, when associated

**Table 1** Agroforestry practices associated with policy land use: agriculture, forestry and urban and peri-urban areas

| Land use and agroforestry practice | | Common name | Brief description |
|---|---|---|---|
| AGRICULTURE | Silvopasture | Wood pasture and parkland | Typically areas of widely spaced trees that are also used for forage and animal production |
| | | Meadow orchards | This practice includes fruit orchards, shrubs which are grazed or sown with pastures, but also olive groves and vineyards |
| | Silvoarable | Hedgerows and windbreak systems | Here the woody components are planted to provide shelter, shade or parcel demarcation to a crop and/or livestock production system |
| | | Alley-cropping systems | Widely spaced woody perennials inter-cropped with annual or perennial crops. It comprises alley cropping, scattered trees and orchards and line belts within the plots. These practices are sometimes found only during the first few years of the plantation |
| | Riparian buffer strips | Riparian buffer strips | Areas of tree and shrubs allowed to establish croplands/pastures and water sources such as streams, lakes, wetlands and ponds to protect water quality can be identified as silvoarable or silvopasture |
| FOREST | Silvopasture | Forest grazing | Although the land cover is described as forest, the understory is grazed |
| | Forest farming | Forest farming | Forested areas used for production or harvest of naturally standing specialty crops for medicinal, ornamental or culinary uses |
| URBAN AND PERI-URBAN | Homegardens | Homegardens | Combining trees/shrubs with vegetable production usually associated with peri-urban or urban areas |

with wetlands or croplands, has been described by Nieto et al. (2014) as a good long-term strategy to promote pollination. When the presence of different species of trees or shrubs is assessed at farm scale, woody perennials have been found to enlarge and complement the overall plant flowering period, facilitating pollination and insect abundance that could be profitably exploited (honey production). In addition, the presence of trees and shrubs reduces water pollution in aquatic environments where good water quality is of paramount importance for the survival of aquatic species as well as birds and other animals. At landscape level, the presence of woody perennials in different farms enhances the impact that trees or shrubs can have on biodiversity inside and outside the farm by modifying the environment through wind speed reduction, as well as affecting water cycling and therefore increasing micro-habitat heterogeneity that contributes to biodiversity promotion. The presence of woody perennials also positively affects soil microorganisms, insect and vascular plants, acting in a holistic way through the whole ecosystem.

When agroforestry is implemented in permanent grasslands, differences in plant species composition between the area below the tree and the open areas occur, due to the higher level of nutrients below trees, as a result of leaf litter accumulation, favouring the presence of monocots over that of dicots, including legumes (Pardini and Rigueiro-Rodríguez 2010). Permanent grasslands including woody perennials support livestock production, providing a source of feed by enlarging the feeding season from grasses which continue to grow below the tree when grasses cease to grow in open sites (Fig. 1), or by directly providing feed through their leaves/stems or their fruits. Europe hosts half of the autochthonous livestock breeds, half of which are at risk of extinction (Rois et al. 2006), and these breeds find their ideal habitat in permanent grasslands with woody perennials, highlighting their role in animal biodiversity conservation.

The same advantages apply where fruit trees are part of the system and herbaceous vegetation is used by livestock. Through keeping grassy

**Figure 1** Tree shade effect on below tree vegetation at the end of the growing season in the dehesa (Spanish agroforestry system where low tree (*Quercus ilex* and *Q. suber*) densities are combined with agricultural and/or pastoral activities), promoting the extension of the growing season to feed animals. Photo: Gerardo Moreno.

areas under control (reducing weeds and potential insect pest habitats), sheep grazing has been described as a good practice to reduce the need of pesticides for some fruit trees such as chestnuts, maintaining biodiversity across different scales. When silvopastoralism is used in permanent grasslands with woody perennials, permanent crops and forestry, the presence of animals enhances biodiversity through the creation of microsites caused by trampling, faeces deposition and plant species selection made by animals (Buttler et al. 2009). All these features increase niche heterogeneity and therefore biodiversity. The use of grazing animals in forestland where fire risk is high is also one of the cheapest solutions to reduce fire risk and consequently the risk of biodiversity losses (Etienne 1996; Castro 2009; Casals et al. 2009; Papanastasis et al. 2009).

## 3 Agroforestry and the protection of species and habitats

Due to the large proportion of degraded land in Europe, the European Commission has issued directives aiming at protecting habitats that still preserve biodiversity and provide more public goods than degraded and unprotected lands (EC 2011a and b). The Habitat and Birds Directives linked to the Nature 2000 network, together with the high nature value (HNV) systems, are examples of key EU biodiversity protection initiatives. The Nature 2000 Network represents 18% of the EU's land area, being the largest protected area network in the world (EC 2009), while high nature value (HNV) systems are estimated to cover as much as 30% of EU-28 agricultural land (Keenleyside et al. 2014). The first target of the EU biodiversity strategy aims at protecting species and habitats through the EU NATURA 2000 network and aims at better conservation or a secure status for 100% more habitats and 50% more species by the year 2020.

There are four key EU actions for the protection of species and habitats:

(i) the completion of the Natura 2000 Network (already achieved) (EC 2009);
(ii) ensuring its adequate management and funding;
(iii) monitoring and reporting biodiversity through ICT tools at European level; and
(iv) involving society through the increasing biodiversity awareness.

Due to the important role that woody perennials and agroforestry play in biodiversity promotion around Europe, this monitoring should specifically look at the presence of agroforestry practices in the European landscape. A comparison between the map of the higher nature value (HNV) farmland area in Europe (EEA 2012) and the map of the areas allocated to silvopastoralism reveals that most silvopastoral areas are located in the Mediterranean area, and

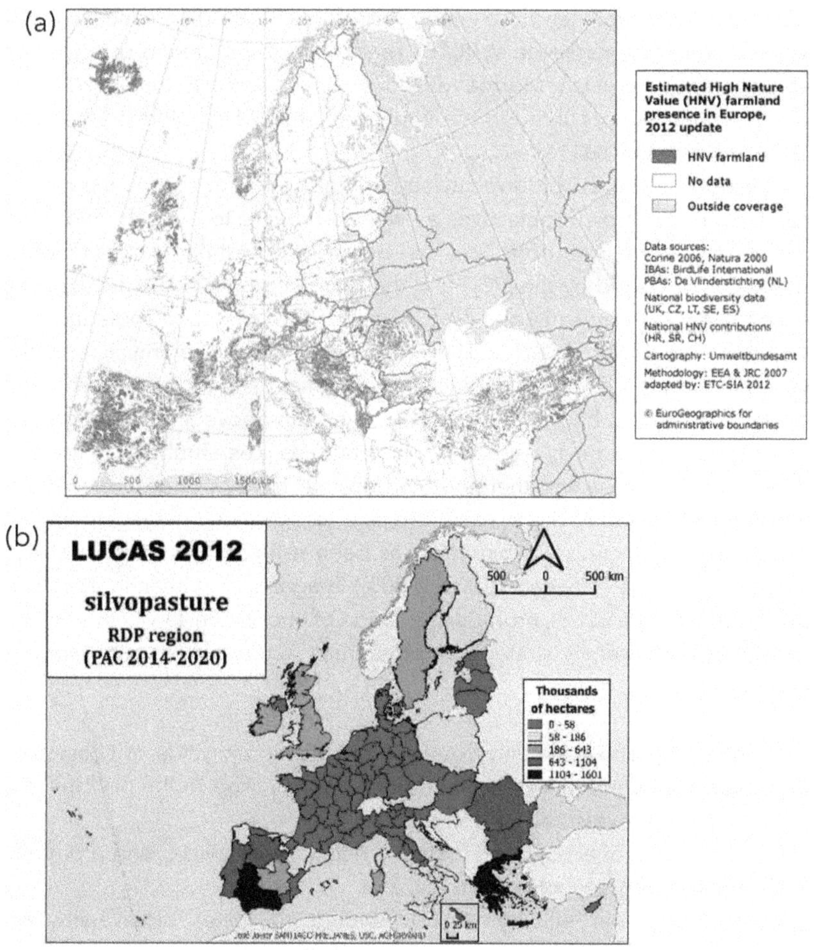


**Figure 2** (a) Estimated high nature value (HNV) farmland area in Europe (EEA 2012) and (b) the silvopasture area in Europe (Mosquera-Losada et al. 2018a).

that most of the HNV farmland is managed following agroforestry principles (Fig. 2).

## 4 Agroforestry and the maintenance and restoration of ecosystems

Target 2 of the EU biodiversity strategy (EC 2011b) aims at maintaining and restoring ecosystems and their services by including green infrastructures in spatial planning and restoring at least 15% of degraded ecosystems by

2020. As a sustainable land-use system, agroforestry could be used to restore degraded ecosystems (Kay et al. 2018). There are good examples of ecosystem services restoration in lands formerly used for mining and in mountain areas that have been carried out through the implementation of agroforestry practices (Quinkenstein et al. 2011).

Target 2 of the EU biodiversity strategy also aims at contributing to the EU's sustainable growth objectives as well as helping to mitigate and adapt to climate change. Both objectives can be achieved with agroforestry, which has been recognized by the IPCC and the EU (Decision 529/2013/EU) as one of the most important strategies for climate change mitigation and adaptation (Mosquera-Losada et al. 2017). The presence of woody perennials in arable systems makes it possible to produce more biomass and therefore carbon per unit of land that can be stored for a long time both above and below ground. Below ground storage is especially important, as tree/shrub roots are able to introduce biomass and therefore carbon into the soil at a deeper depth, reducing $CO_2$ losses to the atmosphere as a result of land management (e.g. ploughing). The increase of biomass has been estimated to be at least 40% more (with a range between 20% and 80%) in agroforestry than in open sites (Dupraz and Liagre 2008), promoting a range of ecosystem services.

The EU biodiversity strategy includes three actions in order to tackle its second target:

- (i) mapping and assessing the state and economic value of ecosystems and their services in the entire EU territory, linked to the principle that 'what gets measured gets saved';
- (ii) restoration of ecosystems fostering ecosystem services, and promoting the use of green infrastructures; and
- (iii) assessing the efficacy of EU funding to foster biodiversity, and developing schemes to pay for ecosystem services.

Measurements of agroforestry land have been carried out by den Herder et al. (2017) and Mosquera-Losada et al. (2018a) looking respectively at only trees and trees or/and shrubs. Kay et al. (2018) has provided a methodology to measure the state of ecosystem services in Europe, associating agroforestry with those areas where there is a higher need to enhance ecosystem services. It identifies the positive influence of agroforestry systems on the supply of regulating services and their role in enhancing landscape structure. Mosquera-Losada et al. (2018b) have evaluated the effects of the main funding agriculture scheme in Europe, the common agriculture policy (CAP), on agroforestry, highlighting that agroforestry is promoted by the EU but not always recognized as such. As an example, homegarden agroforestry practices should be considered within the concept of smart-villages trying to better connect rural and urban areas. As a result they are not usually directly funded by the CAP, as urban areas are out of the scope of the CAP.

## 5 Achieving more sustainable agriculture and forestry

Agroforestry is linked to integration of agriculture and forestry at landscape level (Mosquera-Losada 2018a; Santiago-Freijanes et al. 2018a). By introducing woody perennials in agricultural lands or agricultural activities in forest lands and connecting both types of land use (green corridors connecting forest and agricultural areas), biodiversity is enhanced. This can be done through the use of understory forage in forest land when forage availability is scarce in open grasslands. The connection between forest and open grasslands through the involvement of livestock has a long tradition and can involve short (transtermitance) or long distances (transhumance) (Bunce et al. 2009). Moreover, the existence of scattered trees in arable lands can be an excellent corridor connecting forestland biodiversity (large mammals and birds). Sustainable agriculture and forestry should be linked to the use of the relevant resources, either the understory in forests or the trees or shrubs in agricultural lands. The recently established European bioeconomy strategy aims at connecting these two types of land use as a source of goods and services.

Management of shrub hedges is time-consuming but helps to reduce the effect of winds on croplands while delivering products (usually biomass) that can, for example, be used as fuel for heating. Some animals can also consume the shrubs during periods of shortage of grassland, contributing to increase the resilience of livestock farming systems. Shrubs can also be used as a source of nutrients for the soil. In Galicia, for example, legume shrubs are traditionally sown to increase soil fertility and, after processing, as fertilizer. The most used legume species was *Ulex europaeus* with a productivity that can be over 100 Mg DM ha$^{-1}$ after 8 years. Plants were harvested and used as animal bedding which, when enriched with faeces and urine, reduced the C/N relationship, and therefore improved the release of nutrients when applied to arable lands. Improvement of nutrients also increases biodiversity through the application of organic matter into those soils that have a low organic-matter content.

Target 3 of the EU biodiversity strategy includes 5 actions (from action 8 to action 12 of the strategy) to which agroforestry can contribute. Action 8 is related to the enhancement of CAP direct payments to reward environmental public goods such as crop rotation and permanent pastures. The strategy also allows the improvement of cross-compliance standards for GAEC (good agricultural and environmental conditions), by also including the provisions of the Water Framework directive. International and European policies dealing with agroforestry have been described by Mosquera-Losada et al. (2018a) and Santiago-Freijanes et al. (2018b), while excellent recommendations to foster agroforestry across the world has been reviewed by Buttoud (2013). CAP payments to agroforestry have been included in the last (2007-2013) and current (2014–2020) CAP with different degree of success, where agroforestry is promoted through different policy measures. The next CAP (2021-2027) aims at providing funds based on evidence or result-based payments linked to the provision of ecosystem services. Agroforestry is associated with GAEC as agroforestry practices foster hedgerows and trees (Santiago-Freijanes et al. 2018c) that also improve water quality.

Action 9 of the EU biodiversity strategy is related to better orientation of rural development policies towards achieving biodiversity needs and developing tools to help farmers and foresters work together towards biodiversity conservation. The creation of knowledge reservoirs AFINET (http://www.eurafagroforestry.eu/afinet) and handbooks (www.agroforestrynet.eu) integrates knowledge from research projects and farmer practices. Action 10 is related to the conservation and support of genetic diversity in Europe's agriculture that can be linked to the preservation of crop varieties better adapted to shading conditions (Mosquera-Losada et al. 2019) or autochthonous breeds (Rois et al. 2006) as mentioned before.

Action 11 of the EU biodiversity strategy is related to the encouragement of forest owners to protect and enhance forest biodiversity through adequate activities such as agroforestry (Mosquera-Losada et al. 2006) as well as the use of animal breeds that are adapted to forest areas (Mosquera-Losada et al. 2017). This is also related to action 12, related to the promotion of the integration between biodiversity measures such as fire prevention and the preservation of wilderness areas in forest management plans.

## 6 Stopping the loss of global biodiversity

The EU is committed to stepping up its contribution in the fight to protect global biodiversity in line with international commitments made in October 2010, when the UN Convention on Biological Diversity adopted a strategic plan to address global biodiversity loss over the coming decade. This objective is based on four actions (from action 17 to action 20 of the EU biodiversity strategy). Action 17 aims at reducing the impacts of EU consumption patterns on biodiversity and to make sure that the EU initiative on resource efficiency and trade negotiations all reflect this objective. Consumer education and the promotion of extensive use of local food was one of the challenges mentioned by the farmers in the AFINET project to foster the transition from conventional to agroforestry systems. The use of local food and resources is key to maintain biodiversity at both the EU and international scale. Action 18 (target more EU funding towards global biodiversity and make this funding more effective), action 19 (systematically screen EU action for development cooperation to reduce any negative impacts on biodiversity) and action 20 (make sure that the benefits of nature's genetic resources are shared fairly and equitably) are also related with agroforestry.

## 7 Future trends

Future trends in research should focus on the most appropriate woody perennials to be integrated with agricultural production in different environments. Recognition of agroforestry as part of the way to fulfil sustainable development goals should be researched in order to provide standardized indicators that should be suitable to each type of land cover such as forestry, permanent grasslands, permanent crops and arable lands. Silvopasture increases the biodiversity of forest, permanent grasslands and permanent crops. Stocking rate should be the primary factor to evaluate the effect of livestock on both biodiversity and woody perennials. Evaluation of arable crops biodiversity should be conducted to find the most suitable arable crops varieties that are able to grow under the different types of shading considering light intensity and tree canopy cover. Special attention should be given to the evaluation of the advantages of agroforestry in HNV areas where biodiversity should be protected. Research on land restoration degraded by different factors such as mining or fires should be specifically targeted due to the large amount of existing degraded areas. Research should focus on the use of pioneer species able to cope with particular types of degraded land. Moreover, the comparison of different woody perennial alternatives in provision of different ecosystem services should be evaluated. CAP agroforestry accountability is key to understand the contribution of this type of land-use system in halting biodiversity losses. More research on agroforestry accountability at CAP, but also at global level, is needed.

## 8 Where to look for further information

Below is a list of papers for more information on agroforestry and biodiversity:

- Fernández-Núñez, E., Rigueiro-Rodríguez, A. and Mosquera-Losada, M. R. 2014. Silvopastoral systems established with Pinus radiata D. Don and Betula pubescens Ehrh.: tree growth, understorey biomass and vascular plant biodiversity. Forestry 87, 512–24.
- Ferreiro-Domínguez, N., Rigueiro-Rodríguez, A. and Mosquera-Losada, M. R. 2011. Response to sewage sludge fertilisation in a Quercus rubra L. silvopastoral system: soil, plant biodiversity and tree and pasture production. Agriculture, Ecosystems and Environment 141, 49-57.

- Mosquera-Losada, M. R., Rodríguez-Barreira, S., López-Díaz, M. L., Fernández-Núñez, E. and Rigueiro-Rodríguez, A. 2009. Biodiversity and silvopastoral system use change in very acid soils. Agriculture, Ecosystems and Environment 131, 315–24.
- Rigueiro-Rodríguez, A., Ferreiro-Domínguez, N. and Mosquera-Losada, M. R. 2010. The effects of fertilization with anaerobic, composted and pelletized sewage sludge on soil, tree growth, pasture production and biodiversity in a silvopastoral system under ash (Fraxinus excelsior L). Grass and Forage 65, 248–59.
- Rigueiro-Rodríguez, A., Mouhbi, R., Santiago-Freijanes, J. J., González-Hernández, M. P. and Mosquera-Losada, M. R. 2012. Horse grazing systems: understory biomass and plant biodiversity of a Pinus radiata stand. Scientia Agricola 69(1).

## 9 Acknowledgements

This chapter was partially carried out under the H2020 AFINET project: Agroforestry Innovation Networks; Grant agreement 727872. This work was undertaken by the University of Santiago de Compostela (USC) to support the objectives of the Global Research Alliance on agricultural greenhouse gases (www.globalresearchalliance.org). The information contained within should not be taken to represent the views of the Alliance as a whole or its partners.

## 10 References

Bunce, R. G. H., Pérez-Soba, M. and Smith, M. 2009. *Assessment of the Extent of Agroforestry Systems in Europe and Their Role Whiting Transhumance Systems*, pp. 321–42.

Buttler, A., Kohler, F. and Gillet, F. 2009. The Swiss mountain wooded pastures: patterns and processes. In: Rigueiro-Rodriguez, A., McAdam, J. and Mosquera-Losada, M. R. (Eds.), *Agroforestry in Europe*. Springer, Dordrecht, pp. 377–96.

Buttoud, G. 2013. *Advancing Agroforestry on the Policy Agenda*. FAO. http://www.fao.org/3/a-i3182e.pdf (accessed 05 May 2019).

Casals, T., Baiges, T., Bota, G., Chocarro, C., Bello, F., Fanlo, R., Sebastiá, M. T. and Taull, M. 2009. Silvopastoral systems in the North-eastern Iberian Peninsula: a multifunctional perspective. In: Rigueiro-Rodriguez, A., McAdam, J. and Mosquera-Losada, M. R. (Eds.), *Agroforestry in Europe*. Springer, Dordrecht, pp. 377–96.

Castro, M. 2009. Silvopastoal systems in Portugal: current status and future prospects. In: Rigueiro-Rodriguez, A., McAdam, J. and Mosquera-Losada, M. R. (Eds.), *Agroforestry in Europe*. Springer, Dordrecht, pp. 161–82.

CBD 2010. Strategic plan for biodiversity 2011–2020, including Aichi targets. Available at: https://www.cbd.int/sp/default.shtml (accessed 05 May 2019).

Den Herder, M., Moreno, G., Mosquera-Losada, R. M., Palma, J. H. N., Sidiropoulou, A., Santiago-Freijanes, J. J., Crous-Duran, J., Paulo, J. A., Tomé, M., Pantera, A.,

Papanastasis, V. P., Mantzanas, K., Pachana, P., Papadopoulos, A., Plieninger, T. and Burgess, P. J. 2017. Current extent and stratification of agroforestry in the European Union. *Agriculture, Ecosystems and Environment* 241, 121–32. doi:10.1016/j.agee.2017.03.005.

Dupraz, C. and Liagre, F. 2008. *Agroforesterue: des arbres et des cultures. France Agricole*.

EC 2009. *The EU's Protected Areas – Natura 2000*. Available at: http://ec.europa.eu/environment/basics/natural-capital/natura2000/index_en.htm (accessed 05 May 2019).

EC 2011a. Communication from the commission to the European Parliament, the Council, the economic and social committee and the committee of the regions. Our life insurance, our natural capital: an EU biodiversity strategy to 2020. Available at: https://eur-lex.europa.eu/legal-content/EN/TXT/PDF/?uri=CELEX:52011DC0244&from=EN (accessed 05 May 2019).

EC 2011b. The EU biodiversity Strategy to 2020. Available at: http://ec.europa.eu/environment/nature/info/pubs/docs/brochures/2020%20Biod%20brochure%20final%20lowres.pdf (accessed 05 May 2019).

EEA 2012. Estimated HNV farmland presence in Europe. Available at: https://www.eea.europa.eu/data-and-maps/figures/estimated-high-nature-hnv-presence (accessed 05 May 2019).

Etienne, M. 1996. *Western European Silvopasatoral Systems. INRA Editions*.

Kay, S., Crous-Duran, J., Ferreiro-Domínguez, N., García de Jalón, S., Graves, A., Moreno, G., Mosquera-Losada, M. R., Palma, J. H. N., Roces-Díaz, J. V., Santiago-Freijanes, J. J., Szerencsits, E., Weibel, R. and Herzog, F. 2018. Spatial similarities between European agroforestry systems and ecosystem services at the landscape scale. *Agroforestry Systems* 92(4), 1075–89. (doi:10.1007/s10457-017-0132-3).

Keenleyside, C., Beaufoy, G., Tucker, G. and Jones, G. 2014. High Nature Value farming throughout EU-27 and its financial support under the CAP. Institute for European Environmental Policy. Available at: http://ec.europa.eu/environment/agriculture/pdf/High%20Nature%20Value%20farming.pdf (accessed 05 May 2019).

Maes 2017. A model for the assessment of habitat conservation status in the EU. JRC Scientific and Policy reports. Available at: https://core.ac.uk/download/pdf/38626714.pdf (accessed 05 May 2019).

Mosquera-Losada, M. R., McAdam, J. and Rigueiro-Rodríguez, A. 2006. *Silvopastoralism and Sustainable Land Management*. CAB International, Wallingford.

Mosquera-Losada, M. R., Santiago Freijanes, J. J., Pisanelli, A., Rois, M., Smith, J., den Herder, M., Moreno, G., Malignier, N., Mirazo, J. R., Lamersdorf, N., Ferreiro Domínguez, N., Balaguer, F., Pantera, A., Rigueiro-Rodríguez, A., Gonzalez-Hernández, P., Fernández-Lorenzo, J. L., Romero-Franco, R., Chalmin, A., Garcia de Jalon, S., Garnett, K., Graves, A. and Burgess, P. J. 2016. Extent and success of current policy measures to promote agroforestry across Europe. Deliverable 8.23 for EU FP7 Research Project: AGFORWARD 613520, 95 pp. Available at: https://www.agforward.eu/index.php/es/extent-and-success-of-current-policy-measures-to-promote-agroforestry-across-europe.html?file=files/agforward/documents/Deliverable%208.23%20Extent%20and%20Success%20of%20Current%20Policy%20Measures%208%20Dec%202016.pdf (accessed 05 May 2019).

Mosquera-Losada, M. R., Borek, R., Balaguer, F. and Mezzarila, G. and Ramos-Font, M. E. 2017. *Agroforestry as a Mitigation and Adaptation Tool*. Available at: https://ec.euro

pa.eu/eip/agriculture/sites/agri-eip/files/fg22_mp9_cc_adaptation_mitigation_2017_en.pdf (accessed 05 May 2019).

Mosquera-Losada, M. R., Santiago-Freijanes, J. J., Rois-Díaz, M., Moreno, G., den Herder, M., Aldrey-Vázquez, J. A., Ferreiro-Domínguez, N., Pantera, A., Pisanelli, A. and Rigueiro-Rodríguez, A. 2018a. Agroforestry in Europe: a land management policy tool to combat climate change. *Land Use Policy* 78, 603–13. (doi:10.1016/j.landusepol.2018.06.052).

Mosquera-Losada, M. R., Santiago-Freijanes, J. J., Pisanelli, A., Rois, M., Smith, J., den Herder, M., Moreno, G., Ferreiro-Domínguez, N., Malignier, N., Lamersdorf, N., Balaguer, F., Pantera, A., Rigueiro-Rodríguez, A., Aldrey, J. A., Gonzalez-Hernández, M. P., Fernández-Lorenzo, J. L., Romero-Franco, R. and Burgess, P. J. 2018b. Agroforestry in the European common agricultural policy. *Agroforestry Systems* 92(4), 1117–27. (doi:10.1007/s10457-018-0251-5).

Mosquera-Losada, M. R., López-Díaz, M. L., Ferreiro-Domínguez, N., Rodríguez-Rigueiro, F. J., Arias Martínez, D., Santiago-Freijanes, J. J., Coello, J., Papadopoulou, P. and Rigueiro-Rodríguez, A. 2019. Silvoarable practices as a mechanism to enhance climate adaptation. North American Agroforestry conference. Agroforestry for sustainable production and resilient landscapes. AFTA 2019 Conference. https://afta2019.org/program/ (accessed 05 May 2019).

Nieto, A., Roberts, S. P. M., Kemp, J., Rasmont, P., Kuhlmann, M., Criado, M. G., Biesmeijer, J. C., Bogusch, P., Dathe, H. H., De la Rúa, P., De Meulemeester, T., Dehon, M., Dewulf, A., Ortiz-Sánchez, F. J., Lhomme, P., Pauly, A., Potts, S. G., Praz, C., Quaranta, M., Radchenko, V. G., Scheuchl, E., Smit, J., Straka, J., Terzo, M., Tomozii, B., Window, J. and Michez, D. 2014. European Red List of Bees, European Commission. https://ec.europa.eu/environment/nature/conservation/species/redlist/downloads/European_bees.pdf

Papanastasis, V. P., Mantzanas, K., Dini-Papanastasi, O. and Ispikoudis, I. 2009. Traditional agroforestry systems and their evolution in Greece. In: Rigueiro-Rodríguez, A., McAdam, J. and Mosquera-Losada, M. R. (Eds.), *Agroforestry in Europe*. Springer, Dordrecht, pp. 89–109. doi:10.1007/978-1-4020-8272-6_5.

Pardini, Mori S. and Rigueiro-Rodríguez, M.-L. 2010. *Efecto del arbolado en la producción de pasto y trigo ("Triticum aestivum") ecológicos en la maremma toscana* (Italia central). *Revista de la Sociedad Española para el Estudio de los Pastos* 40(2), 211–33.

Quinkenstein, A., Böhm, C., Silva Matos, E., Freese, D. and Hüttl, R. 2011. Assessing the carbon sequestration in short rotation coppices of Robinia pseudoacacia L. on marginal sites in Northeast Germany. In: Kumar, B. M. and Nair, P. K. R. (Eds.), *Carbon Sequestration Potential of Agroforestry Systems: Opportunities and Challenges*. Springer, pp. 201–16.

Rigueiro-Rodriguez, A., McAdam, J. and Mosquera-Losada, M. R. 2009. *Agroforestry in Europe*. Springer, Dordrecht.

Rigueiro-Rodríguez, A., Rois-Díaz, M. and Mosquera-Losada, M. R. 2010. Integrating silvopastoralism and biodiversity conservation. In: Lichfouse, E. (Ed.), *Biodiversity, Biofuels, Agroforestry and Conservsation Agriculture*. Springer, pp. 359–73. doi:10.1007/978-90-481-9513-8_12.

Rois-Díaz, M., Mosquera-Losada, M. R. and Rigueiro-Rodríguez, A. 2006. Biodiversity Indicators on Silvopastoralism across Eruope. European Forest Institute Technical report 21, 66 pp. Available at: file:///C:/Users/Usuario/Downloads/AS3452428396011521459323715063_content_1%20(1).pdf (accessed 05 May 2019).

Santiago-Freijanes, J. J., Mosquera-Losada, M. R., Rois-Díaz, M., Ferreiro-Domínguez, N., Pantera, A., Aldrey, J. A. and Rigueiro-Rodríguez, A. 2018a. Global and European policies to foster agricultural sustainability: agroforestry. *Agroforestry Systems*, 1–16. doi:10.1007/s10457-018-0215-9.

Santiago-Freijanes, J. J., Pisanelli, A., Rois-Díaz, M., Aldrey-Vázquez, J. A., Rigueiro-Rodrígueza, A., Pantera, A., Vity, A., Lojka, B., Ferreiro-Domínguez, N. and Mosquera-Losada, M. R. 2018b. Agroforestry development in Europe: policy issues. *Land Use Policy* 76, 144–56. doi:10.1016/j.landusepol.2018.03.014.

Santiago-Freijanes, J. J., Rigueiro-Rodríguez, A., Aldrey, J. A., Moreno, G., den Herder, M., Burgess, P. and Mosquera-Losada, M. R. 2018c. Understanding Agroforestry practices in Europe through landscape features policy promotion. *Agroforestry Systems* 92(4), 1105–115. doi:10.1007/s10457-018-0212-z.

UN 1992. The Rio declaration on environment and development. Available at: http://www.unesco.org/education/pdf/RIO_E.PDF (accessed 05 May 2019).

www.ingramcontent.com/pod-product-compliance
Lightning Source LLC
Jackson TN
JSHW050736261224
76011JS00005B/38

* 9 7 8 1 8 0 1 4 6 9 8 9 0 *